Universal Darwinism:

The path of knowledge

John Campbell

Introduction

The major theme of this book is the claim that information and knowledge are prerequisite for physical existence, not only for the physical existence of those entities described by fundamental physics but also for the emergent entities described by chemistry, biology and culture. The reason knowledge is prerequisite for the existence of physical entities is because they are complex and the second law of thermodynamics forbids the existence of complexity except in extremely rare and contrived situations. Finding and maintaining such situations does not happen by accident; it requires knowledge. I will argue that the knowledge required for existence is accumulated in accord with the mathematics of Bayesian inference and that the singular known method of instantiating inference within physical reality is through the operation of Darwinian processes.

I will cite the scientific literature for instances where knowledge accumulated through Darwinian processes is used to describe the evolution of complex, low entropy entities. Such knowledge repositories include the wave function of quantum entities, the genetics of life forms, the neural connections of animals and the literature of science.

This argument is motivation for the theory of Universal Darwinism. The Universal Darwinism meta-theory contains the numerous scientific theories which employ a Darwinian process to explain the creation and evolution of their subject matter as well as an exposition of the general principles these theories have in common. The 'universal' aspect of this theory is justified by the broad scope of subject matter included under its umbrella. The literature contains numerous scientific theories in quantum physics, atomic and molecular physics, cosmology, biology and culture. We will examine a number of these theories in some detail.

Over the past several hundred years a hallmark of our evolving scientific understanding has been the ability to progressively explain wider portions of reality within the scope of individual theories. For example Newton's universal gravitation explained both the motions of terrestrial objects such as falling apples and the motion of planetary bodies within the solar system. More recently the standard model of particle physics has united electromagnetism with forces operating within atoms to provide a complete theory of the micro-world. Current work on gravity (Verlinde 2010) indicates it may be best explained as a force emerging from those same micro-world forces, offering the tantalizing prospect that all fundamental interactions within physical reality may soon be explained within the scope of a single theory.

The progress made by scientific theories to progressively explain greater scopes of subject matter has not just occurred in physics. The merging of Darwin's theory of evolution with understanding of DNA has led to a central explanatory theory for biology as has plate tectonics for geology.

This book will make the argument that Universal Darwinism provides a further advance in the unification of scientific understanding; that Universal Darwinism is a means of consolidating a wide swath of seemingly disparate scientific subject matter within a single theoretical paradigm.

The forces and interactions of the micro-world are viewed by modern particle physics in terms of information. John Archibald Wheeler, one of the past century's most influential physicists, liked to say that his career had moved through three phases, from "Everything is particles" to "Everything is fields" to "Everything is information" (Ford 2010).

The concept of information not only has a central explanatory role within particle physics it is also central to explanations of emergent levels of complex reality such as biology and culture. The increased focus of scientific explanation on information holds promise for a single theory with the ability to unite many branches of science within the same theoretical framework.

For the past 60 years a single theory that is able to explain all fundamental interaction in terms of information has been physic's Holy Grail and is often referred to as the Theory of Everything (TOE). The presumptuous title is due to the hope that given a single theory of the building blocks we would be able to explain all the more complex emergent systems, such as chemistry and life, which arise from them.

While this ambitious goal might seem possible in principle it rapidly runs into a number of practical problems. First, trying to make predictions concerning more complex systems in terms of fundamental physical theories has not been practically possible. The complex calculations involved with trying to describe even atoms as simple as helium in terms of fundamental particle physics have proved unsolvable. Emergent systems such as chemistry and life are most easily and naturally understood using their own emergent laws in addition to those of physics. It has become clear, to most researchers, that a true Theory of Everything must integrate the laws of these emergent fields with those of physics.

Although information is a single unifying concept which may be employed in the description of much of scientific subject matter, this fundamental conceptual nature might not imply a fundamental physical nature. This is a common explanatory paradox, where concepts which provide a simple explanatory framework may not be simple in themselves. For example, although atoms and their isotopes may be simply conceived as only different combinations of protons and neutrons we may not assume that protons and neutrons are physically simple or fundamental. Similarly, although much of genetics may be simply conceived in terms of the DNA base pairs T, A, G and C, these base pairs are themselves complex and do not have a simple physical nature. Likewise information, although conceptually simple, has in fact a quite complicated structure.

Information's scientific definition is in terms of expectations and probability. It is the amount of surprise that comes with an experience. If an experience is unexpected or surprising then a lot of information is received. If it was largely expected then little information was received.

While complex emergent levels of reality may be described using information as a simple building block it is clear that a deeper understanding will entail taking the complex nature of information fully into account. Any integrated theory capable of explaining both the reality of fundamental physics and the further layers of complexity emergent from it will have to contend with this fact.

This book explores a path, already broadly sketched in the scientific literature, by which nature has utilized information mechanisms first to manifest a classical reality from the quantum substrate and then to create emergent designs capable of supporting further layers of complex emergent structures. I argue that complex entities must contain specially adapted information or knowledge mechanisms and that these have been physically selected and accumulated by methods equivalent to the Darwinian process.

Information is not a fundamental concept in the sense of being self-explanatory; it requires further explanation. Its scientific definition in terms of probability, expectations and surprise imply sophisticated underlying machinery. Implicit in the meaning of expectation or surprise is the comparison of two representations: one received through experience versus a model of expectations. In order to receive information it appears necessary for a system to possess an internal model of expected phenomena which can be compared to actual experience.

Although requiring information systems to possess an internal model adds complexity, it may also provide a way forward in understanding the emergence and evolution of information systems over evolutionary time. Defining knowledge as those attributes of an internal model which serve to reduce the surprise encountered when experiencing actual outcomes provides a mathematically tractable method of examining knowledge as it has evolved in emergent systems. In order for an internal model to contain knowledge or to reduce surprise when information is received the model must be updated or refined by the information it does receive. A mathematical theorem requires that the updating of any model to more accurately reflect experience must take place according to the rules of Bayesian probability (Jaynes 2003).

Scientific understanding has come to regard objective reality as a 'point of view invariant' realm where information that can be known about any given entity by a second entity or observer must be consistent with what may be known by a third. In order to be objective, scientific law must provide consistent explanations and predictions for each observer or entity regardless of its circumstances (Stenger 2006). The single theory of fundamental forces we are on the brink of uncovering conforms to these standards of objectivity. It describes a single experience for all observers and entities. The evolution of our understanding of objective reality is reviewed in Section 1.

The path of knowledge, which I attempt to document, has been traversed over the course of evolutionary time since the beginnings of the universe in many different guises but all utilizing a common information processing mechanism. Many of these guises have been described in the scientific literature and their information processing mechanism identified as Darwinian processes. Section 2 examines the common components which nature has used to forge complex emergent structures from the web of fundamental particles.

Our best physical theories tell us that one entity within objective reality may influence, interact with, or be detected by another entity only through one of the three fundamental physical forces: electro-weak, strong and gravitational forces. This formulation suggests boundaries delineating a physical arena of objective reality; only those phenomena capable of interactions through one or more of these three fundamental mechanisms exist within objective reality. Our best theories of these three interactions consistently explain them in terms of quantum information transfer. Section 2 includes a review of fundamental forces with emphasis on some recent developments which may bring clarity to the confusing nature of this subject matter.

Recent research by Wojciech Zurek (2009) and others provides a revolutionary new understanding of the manner in which classical reality emerges from quantum processes. Perhaps surprisingly this explanation describes a Darwinian process transferring information from quantum processes to classical reality. This information network forms the web of reality we are a part of and which we experience.

Zurek's findings transform our understanding of quantum processes and in my opinion resolve many of the quantum 'mysteries'. Since the time of Einstein quantum processes have been described even by those responsible for developing them as weird or spooky. When Zurek's interpretation is combined with Information Theory we can see in contrast that quantum processes are in fact quintessential inferential and rational processes.

In fact the outcomes of quantum processes are strictly rational; they are equivalent to computations, quantum computation. A quantum computation considers all possible answers and then chooses one as the actual answer. All quantum interactions involve such computations where the answer to the computation is the outcome of the interaction. In other words the physical future taken by a quantum system is in accordance with a computation.

It is axiomatic in quantum theory that the quantum wave function is the source of all knowledge concerning a quantum process. This wave function evolves over time as a function of the environmental forces acting on it.

When a quantum system is about to interact with another entity in its environment the wave functions of the two entities are said to become entangled and their joint wave function forms a probability distribution assigning a probability to all possible outcomes of the interaction.

Once the interaction has taken place the probability distribution is updated; the probability of the actual outcome becomes 1 and the probability of all other potential outcomes is zero. This means that a second identical interaction taking place immediately after the first will provide the identical outcome with probability 1. With the passing of time the predictive ability of the wave function tends to diminish from certainty, other outcomes gain greater probability and future outcomes become less certain.

This is exactly in accord with the mathematical process of Bayesian Inference where knowledge is gained from experience or evidence. Inference involves a probability distribution over all possible hypotheses involving the outcome of a random event. The experience of the actual outcome is used to update the probability distribution. Roughly those hypotheses predictive of the actual outcome receive greater probability; those predicting something other than the actual outcome receive a reduced probability. In this manner predictive ability is enhanced.

Compounding the perceived weirdness of quantum processes has been the assumption, often enshrined within an axiom, that the wave function has no physical instantiation; it has been considered a purely mathematical aid in calculating outcomes. However as quantum processes themselves are now seen to utilize something equivalent to the wave function in determining their own future and as Zurek has demonstrated there can be no information without physical representation, we must consider it plausible that the wave function assumes a physical structure within quantum systems. Even though science is currently unable to directly probe the physical structure of the wave function the fact that there are over twenty orders of magnitude in scale between the Planck scale, at the basis of quantum processes, and the proton leaves a great deal of room for such structures to exist beneath the level at which science is currently able to probe experimentally.

The process by which the particular outcome of a quantum interaction is selected and the probability distribution inherent in the wave function updated has been described by Zurek. He has found that this process, commonly called quantum decoherence, takes the form of a Darwinian process.

Zurek's view is that during quantum interactions decoherence takes place, which involves an attempt to copy all possible outcomes into classical reality, but the vast majority of these outcomes are unable to survive for an appreciable time in the classical environment. After an extremely short time those outcomes suited for survival are all that exist. This description perfectly meets the criteria of the classic 'copy with selective retention' algorithm defining a Darwinian process. The outcomes selected by decoherence to participate in physical reality are mediated by a Darwinian process which selects information adapted for survival in the classical reality from the vast number of potential quantum outcomes. Zurek has named this process Quantum Darwinism and it effectively forms the boundary of objective or classical reality (2007).

Zurek's work reveals an inherent knowledge mechanism in the form of the quantum wave function which serves to model the environment of quantum entities and the information they may potentially exchange. It may seem unexpected that this knowledge mechanism is very similar to those utilized by other types of complex organization emergent in the web or reality.

From our understanding of the history of the universe it is clear that near its beginning the web of quantum interactions between fundamental particles composed the extent of complexity existent in the universe. Further emergent levels of complexity appeared later forming the subject matter of atomic physics, chemistry, biology, neuroscience and culture. These emergent structures display reduced local entropy suggesting they are highly improbable and that their existence requires explanation.

The principle of Maximum Entropy tells us that systems will evolve to states of lower entropy only when they are constrained to do so by scientific law (Jaynes 2003). Any thorough scientific explanation of the many existing low entropy states should include a statement of the scientific laws responsible for constraining them from visiting higher entropy states.

Many low entropy states in the realm of atomic physics and chemistry are well explained within the framework of Quantum Darwinism by the operation of fundamental forces and particles in a cooling universe and other specialized environments, without the need for additional scientific laws. However, any explanation of the evolution of life on earth and its subsequent evolution into more complex forms does require additional scientific laws.

These explanations and their attendant scientific laws might be understood using the concept of adaptive systems. Adaptive systems, as defined by Karl Friston (2007) and others, involve the dynamics of systems which act to optimize their relationship with their environments in order to retain their low entropy states. Central to

Friston's analysis is the conclusion that adaptive systems must contain models of their environments and of causes within those environments. The optimality of an adaptive system results from its ability to minimize its free energy in an information sense; that is, to reduce the discrepancy or 'surprise' it experiences when interacting with its environment.

Central to this process is the ability of an adaptive system to discover and implement adaptations or structures which mediate the system's interaction with its environment and serve to constrain the system's entropy increase. This process of evolution and discovery is most often accomplished through the operation of a Darwinian process.

In this light much of scientific law responsible for constraining entropy within complex systems might be understood as the design details inherent in the adaptations discovered by adaptive systems that were created and have evolved to their present state through the operation of Darwinian processes (J. Campbell 2009).

A central challenge for our paradigm is to construct explanations, within each emergent field, of the inferential mechanisms which select knowledge from the information present in the environment and accumulate it within internal models.

The unreasonable effectiveness of mathematics in the natural sciences has seemed something of a wonder to many researchers (Wigner 1960). Why is mathematics so powerful in describing natural processes? Max Tegmark (2009) offers a solution, which may appear somewhat obvious, suggesting that physical reality is isomorphic to a mathematical structure which we are gradually discovering.

Bayesian inference has laid claim to be the 'logic of science', that it is the unique mathematical structure providing a valid extension of logic to areas of incomplete knowledge where degrees of plausibility must be dealt with (Jaynes 2003). Science is one such area. However this claim may be extended to all areas of the natural world where structures containing knowledge have evolved.

Information implies the existence of a model which may be understood as a family of hypothesis each of which are assigned a probability measuring the likelihood of a particular outcome of some expected phenomena. Gaining knowledge requires logical inference which operates according to Bayesian principles and involves updating the relative probabilities of the model's hypothesis on the reception of new data. This updating cause the model's hypothesis to be more accurately predictive of data received from the outside world and leads to less expected surprise in the future; that is, it leads to lower entropy. Section 3 reviews the mathematics exploited by nature to forge knowledge from information.

The mathematics of Bayesian inference tells us that there is but a single mathematical mechanism for converting information into knowledge, and that is Bayes' Theorem. However the evolution of knowledge entities within physical reality is not as straightforward as the mathematics. Knowledge entities are by their nature complex, and physical reality in the form of the second law of thermodynamics makes the existence of complex structures exceedingly unlikely. I will argue that physical entities contain inference and knowledge mechanisms precisely for the purpose of neutralizing the destructive effects of the second law.

Section 4 makes the argument that such inference mechanisms operating in physical reality take the form of Darwinian processes. This thesis forms a rationale for the theory of Universal Darwinism which is the collection of scientific theories using the Darwinian process to explain the creation and evolution of their subject matter. Universal Darwinism also suggests that Darwinian processes may be the unique method used to evolve knowledge in reality. Further, I propose that in many instances this process takes the form of an adaptive system containing an internal model which models a strategy for survival within an expected environment. This strategy is knowledgeable of its environment and forms a 'blueprint' for the construction of adaptations which serve to constrain the increase of entropy in accordance with the principle of maximum entropy. These characteristics, in addition to the Darwinian process itself, appear to be implied once Darwinian processes are understood within the context of a knowledge mechanism. Sections 5 through 8 review the evidence and arguments regarding this thesis in the subject matter of atomic physics and chemistry, biology, neuroscience and culture.

It might be apparent that science is the most potent form of knowledge in our cultural repertoire. I will argue that the scientific method, although introduced to culture only a few hundred years ago, is but a rediscovery of the same process of knowledge formation used by nature since the beginning of time. Universal Darwinism attempts to encapsulate this process in its many different guises and an overview of the results of this attempt is provided in Section 9.

Although the mention of Albert Einstein may conjure visions of a pure, abstract scientist, it may be of interest to realize that he also made vital contributions to a number of other branches of human understanding including religious understanding.

Einstein viewed religion as an evolutionary process and predicted that its next stage would be a 'cosmic religious experience' made available largely through an understanding of science. In fact he stated that the purpose of science is to make this cosmic religious experience available to those receptive to it (Einstein 1930).

Given Einstein's immense stature we might ask why this vision of religion is not more widely embraced outside of the scientific community. Why aren't members of the 'Cult of Einstein' handing out pamphlets on the street corners? The short answer may be that science is hard. One must devote oneself wholeheartedly and undertake a huge effort of study in order to understand even a small subset of science. Science has been developed in a largely ad hoc manner where each small piece of subject matter comes with its own idiosyncratic jargon, methods and orientation which may require years of solitary labour to master.

Unlike many conventional religions where one need only decide to take a leap of faith, science requires a commitment to understand reality on the basis of the available evidence.

Universal Darwinism, through offering a simple and consistent explanation across much of scientific subject matter, may make Einstein's cosmic religious experience accessible to greater numbers and serve to 'awaken this feeling and to keep it alive in those who are receptive to it' (Einstein 1930).

Section 10 contains some essays on this issue and more generally on the manner in which Universal Darwinism may inform our worldview.

1: The History of Objective Science

Merriam Webster offers a modern definition of 'objective' (Online 2008):

> *of, relating to, or being an object, phenomenon, or condition in the realm of sensible experience independent of individual thought and perceptible by all observers : having reality independent of the mind*

The idea of an objective reality is a core assumption of science and has provided much guidance for the development of science. A common theme in the development of science has been an expanding tolerance towards objectivity. We are progressively coming to understand, through our scientific knowledge, that we exist in an independent reality that evolved a huge amount of complexity before humans existed and that humans are only a natural extension of this reality and are in no way fundamental to shaping it.

Prior to the beginnings of science (as it is understood in a modern sense) dominant world views were largely those of the great religions, revealed through sacred texts and interpreted by a rigid bureaucracy. For those Europeans who were less theistically inclined Platonic philosophy with its focus on an ideal reality unseen by the senses was an alternative. So too were Aristotelian ideas, eventually morphing into those of the church, that were largely based on a priori ideas concerning the 'natural' state of things and came to be relied upon by reason of authority.

The more 'scientific' ideas, expounded by Archimedes, had little influence through the end of the Middle Ages as only few copies of his works existed. Much of his work was first discovered in relatively recent times.

These sources of knowledge of the reality in which we exist downplayed the ability of the senses to inform our worldview. Justifications in the battle between the Church and Copernican astronomy, culminating in the prosecution of Galileo, hinged on the assertion by the Church that sensory data was not to be relied upon and that the creator could order things as he wished including the reliability of our sensory information. Of course it reserved for itself, on pain of death, the sole right to interpret this capricious divine mind.

With the advent of Newtonian scientific understanding in the 17th century a beautiful picture began to emerge of a heliocentric solar system, accurately predictable from the law of gravitation that applied equally to planets and to objects in our everyday life. This was a giant step towards an understanding that human affairs exist in an objective reality that does not necessarily imbue their situation with special significance. Educated people gained an aesthetically pleasing alternative to the narcissistic view that they were the central focus of creation and that their home planet was at the centre of the universe.

In his twenties Newton became part of a small group of natural philosophers who practised science in a tradition largely established by Francis Bacon. This approach became known as the British Empirical School and stressed the importance of empirical evidence derived from experimentation in support of theoretical explanations. It was a revolutionary change from the 'authority-centric' approach which had dominated since the time of the Greeks and formed the beginnings of scientific enquiry.

Here evidence rather than authority was considered the best guide to truth.

This movement also heralded a revolutionary change in thinking about objectivity. Explanations, in the form of theories, gained support if they in turn were supported by the evidence. Further along this chain the quality of evidence was judged by its repeatability. To hold any weight observations must be repeatable. That is, any individual, given the appropriate setting, must be able to make the same observation as made by any other observer. This consensus amongst a group of individuals concerning the empirical evidence they witness leads to a stronger notion of objectivity and of evidence as an objective fact free from the dictates of authority or subjective interpretation. Here was the beginning of the understanding that objective or ontological reality must look the same to all observers.

This small group of British investigators, practising empirical methods of investigation founded the Royal Society in 1660 and it continues today as one of the world's most prestigious scientific organizations.

Another huge blow to the remaining anthropocentric components of the scientific worldview was dealt by Darwin with his theory of Natural Selection identifying humans as only one more animal form evolved from common ancestors. This theory was seen by many as threatening to human values as it awoke us from our previous shared delusion of being the central purpose of creation. Human beings were revealed to be animals created by the same evolutionary process as all other life forms. Yet Darwin's views were interpreted by some as adding grandeur to our situation. We are revealed as amongst the most complex of living things. Our involvement in cultural entities is unprecedented and sets us apart from other living things. We are joined to the rest of life and to the wider universe in a natural way. Perhaps this theory's greatest gift, was to reveal that the drama of life is most clearly seen in an objective context. A universal process has created us along with all other life forms, it was in full operation long before the existence of humans was even a faint possibility, and it will probably continue long after the human species has ceased to exist. To gain understanding of our true situation we started to abandon the infantile delusion that we are 'special' in the sense of existing outside of natural processes and to be open to this objective world.

Philosophy, up until the late 19th century, had largely been at odds with scientific understanding in general and objectivity in particular. Emmanuel Kant, perhaps the most renowned philosopher of all time, believed that the nature of reality was based on relationships between objects, and that what could be known of objects could only be based on our relationship with them. This excluded the possibility of objective knowledge.

In the mid eighteenth century a new branch of philosophy, the philosophy of science, began to take shape, formed mostly from the work of scientists such as Ernst Mach who viewed science as providing a philosophical vantage point. This work was expanded by the school of Logical Positivists who, during the early 20th century, challenged previous philosophical understanding with the notion that only those things that could be measured should have cognitive significance.

This branch of philosophy had tremendous influence on the thinking of the great scientific works of the early 20th century including the theories of relativity and quantum mechanics.

Einstein revered Mach. He digested Mach's thinking and reaped a huge intellectual harvest. He came to understand that objectivity required constraints on scientific law in the sense that all laws must provide every observer with the same predictions regardless of their own particular circumstances. Special relativity provides common predictions for all observers regardless of their own velocity and General Relativity provides common predictions to all observers regardless of their own acceleration. Einstein was led to General Relativity largely through steadfastness to this principle (Isaacson 2008). The guide of objectivity was so compelling to Einstein that he initially called his theory 'Invariance Theory' and only acquiesced when 'Relativity' became more widely adopted by Max Planck and others (Irving 2009).

Einstein expressed his objectivity requirements by demanding that valid scientific theories must provide all observers with accurate predictions of what they would measure. This might seem quite anthropocentric but he also made it clear that in his view objectivity extended beyond human experience and measurement. General relativity in particular predicts the behaviour of all mass and energy in response to gravity whether or not there is a human observer present.

As the baffling world of the quantum was being probed and a scientific formalism regarding it developed the dominant interpretational model came to centre on the work of Niels Bohr. Bohr understood both the classical world and the quantum world underlying it to be distinct realities. These two realities are spanned by the process of measurement where quantum phenomena emerge in classical reality. He insisted that quantum reality could only be known through measurement and that it did not make sense to talk of quantum reality in terms other that what could be measured. In this sense his theory was epistemological, about what we can know, rather than ontological, about what is really out there.

Einstein, also one of the founders of quantum theory, had a much different point of view and insisted that any fundamental physical theory must describe an objective ontological reality. Philosopher C.P. Snow described the ensuing debate: 'No more profound intellectual debate has ever been conducted' (Isaacson 2008). As described in more detail below, perhaps the most persuasive evidence we have today serves to illuminate how they were both correct, Einstein from an ontological viewpoint and Bohr from an epistemological one.

Some variants of quantum theory have supported an anti-objective interpretation of physical reality culminating in that of Eugene Wigner's (1970) where human consciousness was proposed as a necessary component of quantum measurement and quantum wave collapse. Even the dominant interpretation of quantum theory with its focus on measurement might seem to give special status to humans and their perspective. The paradox resulting in the requirement that a human measurement is necessary for the resolution of quantum reality was brought to the fore by Erwin Schrödinger in his famous thought experiment that has come to be known as the Schrödinger's Cat paradox. This paradox hinges on the assumption, implied by quantum theory, that a cat can neither be considered alive nor dead until its state is measured by a human observer.

The century long confusion over the meaning of quantum mechanics, sometimes even abetted by our leading interpretations of scientific theories concerning the nature of objective physical reality, may recently have been largely resolved (Zurek 2009). The set of axioms underlying quantum theory has been revised and simplified. The most contentious ones (the axiom requiring quantum measurements to produce classical results and the axiom that assigns probabilities for the competing classical results) have been shown to be implied by the other axioms, and are therefore unnecessary as separate axioms. Measurement is no longer seen as a fundamental physical process but rather as just one of many types of physical interaction. At the quantum level all physical interactions between the fundamental building blocks of reality may lead to decoherence, the process where quantum effects are recorded in classical reality. Whether these interactions are viewed by a person and considered measurements is irrelevant. In this sense our new understanding is entirely objective.

A further recent breakthrough in our understanding of objective reality has been made by a direct descendant of Einstein's thinking on 'invariant theory' (Stenger 2006). This work details the assumption that objective reality should be described by scientific laws that have the same form regardless of the circumstances of the observer. If one assumes physical laws must take this 'point of view invariant' form then many of our most basic scientific laws can be derived purely from this assumption including: special relativity, general relativity, quantum gauge theories and the symmetry laws.

That these components of physical law are amongst the most accurate and powerful is strong evidence that reality is objective in the sense of point of view invariance.

In some sense this theory of reality brings us full circle back to Kant. All of reality is a web of relationships. The crucial added ingredient is that this web is objective. No entity in the web has a privileged position. We can have objective knowledge of other entities contained in the web through understanding that our knowledge of other entities is no different than that experienced by any other entity. Our experience of an electron must be the same as a proton's experience of an electron; a valid scientific theory of the electron must describe both experiences; the same law must apply to all situations. It must describe the effects of an electron in an objective reality.

If we accept that reality is at bottom this kind of objective network or web then we must also accept that it has become continuously more complex over time since the 'Big Bang'. It has evolved new hierarchies over time including chemistry, biology and culture. Still none of these newcomers is privileged over the others; each experiences the rest of the web in a common manner.

2: Fundamental principles

Our best physical theories tell us that the objective web of reality is held together by only three types of interactions, the bonds of the three forces of nature: gravity, the electroweak and the strong force. A main thrust of current physical research is an attempt to reduce these three theories to a single 'theory of everything' that operated in undivided splendour near the time of the Big Bang, when things were simpler. Leading candidates in this search are String Theory and Loop Quantum Gravity.

It is clear that the success of these new theories will be dependent on accounting for phenomena occurring at the Planck scale, which is about 10^{-35} meters. When we consider that the diameter of a proton is about 10^{-14} meters we see that there is a greater relative difference in scale between the Planck scale and that of 'fundamental' particles than there is between that of fundamental particles and the everyday phenomena of our experience. We might well expect that as much emergent phenomena will be found to occur below the scale of 'fundamental' particles as occurs at scales above it.

Indeed some physical theories of reality at the Planck scale propose that 'fundamental' particles themselves, such as the photon, may be emergent phenomena (Hamma, et al. 2009). We are thus faced with perhaps a near infinite regress of 'fundamental' realities. It looks like it might be 'turtles all the way down'. In spite of that daunting consideration, we may still be amply rewarded by drawing boundaries around those aspects of reality that have the ability to influence us by interacting directly with us.

The small numbers of interactions described by our physical theories are the only interactions known to exist at the particle physics level. In this sense the Logical Positivists have been exonerated; anything that does not interact by these forces is in principle unknown and unknowable to the web of reality we inhabit and from that vantage point may be said to have no existence.

Although our web of interactions is fundamentally a web of quantum interactions it has, over the history of the universe, evolved emergent structures of great complexity. These structures may also be understood in terms of information or knowledge but this information still interacts through quantum processes and it is this information which is processed and from which inferences are drawn. Quantum processes are the only known processes where information is created and they are solely responsible for injecting information into our web of reality (Lloyd 2007).

As an example, one might certainly balk at the idea that sensory information gathered by biological organisms has much to do with quantum interactions. However when we consider that the eye functions by receiving a sample of light from its environment and that the interaction between these photons from the environment and molecules such as rhodopsin in the retina is quantum interacti1on involving only the transfer of quantum information, we can see that it is precisely these interactions that are the source of any information available to the organism. The eye, the brain and its behavioural reactions function to process this information, draw inferences from it and behave appropriately in response to it. None of these functions, however, are the source of new information concerning the environment.

Likewise hearing depends on the detection of pressure waves. Pressure is transmitted through a quantum interaction where information about molecular momentum is transferred to the environment, including auricular sensors, in a quantum manner.

Perhaps an even more dramatic example involves photosynthesis, the biological process through which the sun's light is converted to a biologically useable form of chemical energy. Photosynthesis is a fundamental process in the web of life but complete details of its operation have long eluded researchers. Particularly baffling has been how this biological process is 95% efficient while our best solar panel technology is about 20% efficient. Recent research indicates the answer may be that nature employs a quantum computation to decide on the most efficient chemical pathways to employ for each photon of light received (Sension 2007). In this process, crucially important to life, a good portion of the underlying quantum information is harnessed to improve efficiency.

We might now see the web of reality as a web of information flow. The information that one entity can have of another composes this web which is bound by the laws of physics and is probabilistic in nature.

The following sections will review aspects of the basic fundamental laws of physics as well as some characteristics of adaptive systems which will be used in later discussions of emergent systems.

Gravitation

Our best physical theory of gravity, reality's dominant macroscopic force, is currently undergoing a most exciting explanatory development. Erik Verlinde (2010) has published a paper that explains gravity in term of information processing. Previously John Archibald Wheeler (1973) had coined a phrase describing the essence of general relativity 'Matter tells space how to curve, and space tells matter how to move'. Still the mechanism of 'telling' was unexplained and we were left with the question: what is the nature of the interaction between space and matter that results in this telling?

Building on the work of 't Hooft and others Verlinde (2010) has shown that gravity emerges naturally from principles of thermodynamics and the holography principle. The holography principle views reality as divided into volumes separated by boundaries or screens. The physics that occurs within a boundary is isomorphic with information encoded on the boundary.

In this view information is stored on surfaces, or screens which divide small volumes, and in this way are natural places to store information about particles that move from one side to the other. Thus we imagine that this information about the location of particles is stored in discrete bits on the screens. The dynamics on each screen is given by some unknown rules, which can be thought of as a way of processing the information that is stored on it.

Wheeler's 'telling' is clarified as an information processing event that transforms information stored on boundaries between spaces into particle dynamics.

Quantum processes

All physical interactions may be viewed as quantum interactions and the theories describing them are the most accurate and powerful in all of science.[1] Thus quantum

---

[1] Gravity is not usually considered a quantum theory but it has been shown that a quantum formulation is equivalent to General Relativity in the classical limit. This is the same limit in which

interactions, at the core of our objective reality, are the gateway through which all of its participants must pass and they form a boundary delineating the extent of reality that can be objectively known. If we are to understand the nature of our objective web we must start with an understanding of quantum mechanics.

Until recently this has not been possible. We have had extremely accurate theories of the quantum for nearly a hundred years but although these theories were useful for calculations unfortunately they shed no light on the nature of the quantum process. As one of the great creators of quantum theory, Richard Feynman said (2009): "*I think I can safely say that nobody understands quantum mechanics.*".

Fortunately this century long impasse has recently been skirted. The work of Wojciech Zurek, of the Los Alamos National Laboratory, and collaborators has revealed in detail the processes of quantum decoherence, the process central to understanding the mysteries of the quantum (Zurek 2007).

Quantum theory may be erected from a number of axioms. Usually an axiomatic system aids explanation and

---

General Relativity has been shown to be valid. It is in trying to extend the theory of Gravity to non-classical situations, including extreme energies and time and space resolutions on the Planck scale, that both formulations fail. As noted in Wikipedia's article on the graviton: 'In this framework, the gravitational interaction is mediated by gravitons, instead of being described in terms of curved spacetime as in general relativity. In the classical limit, both approaches give identical results, which are required to conform to Newton.'

understanding. If a scientific theory can be constructed from a few simple axioms using only the assistance of logic we need only understand the axioms and all the rest of the theory is logically implied.

With Newtonian classical mechanics this is surely the case. If we take Newton's three straightforward laws of motion as axioms, much of the derivative, easily understood, classical mechanics can be extracted using logic alone. Unfortunately this has not been the case with quantum theory. The axioms underlying it are mathematical in nature and do not seem to relate to anything in reality. Worse, they seem contradictory. The following is a set of widely used axioms for quantum theory with comments on problems of understanding they entail.

1) For every physical system there is a corresponding mathematical object called a state vector in Hilbert space that has no physical embodiment. This state vector is the most complete source of information that exists concerning the physical system.

How do we understand this? The most complete information we can have of a physical system is a non-physical mathematical object? Sounds like something outside of our web of reality whose existence is in principle unknowable.

2) The state vector evolves in time according to a continuous, deterministic mathematical function except when a measurement occurs and then it jumps to the state described in 4) and 5) below.

This axiom is a little more promising in that the undisturbed state evolves in a mathematically tractable manner. The jump part seems somewhat contradictory; smooth, continuous evolution and then a jump?

3) Once a measurement is made the state vector assumes a state such that the same measurement immediately reapplied to this state has 100% probability of achieving the previous measured result.

More promising yet, this axiom tells us that although there is a jump involved when we measure a quantum system this jump is not arbitrary and that the system's evolution picks up after the jump at the state revealed by the measurement and resumes its smooth evolution from there.

4) The outcome of any measurement on a physical system can be predicted by performing a specific mathematical operation on its state vector.

This is even more confusing. Predictions can only be made by performing a seemingly arbitrary mathematical operation on the mathematical object that is the source of all knowledge we can have of the quantum system?

5) The outcome of any measurement process on a physical system can only be predicted as a probability for obtaining that result. The procedure for obtaining this probability is known as Born's rule.

Not good either. Predictions can only be made in the form of probabilities and these probabilities must be calculated using another seemingly arbitrary mathematical rule?

A first step we can take towards achieving clarity is to replace the word 'measurement' with 'interaction' in all the axioms above. As guided by our analysis of objective reality we must conclude that interactions conducted by humans and labelled by us 'measurements' are indistinguishable from the same unobserved interaction taking place within the web of objective reality.

Still we are left with a huge lack of clarity that might be summarized:

1) The source of a quantum system's effects that can be experienced by any entity within the web of reality is the wave function, a non-physical mathematical object.

2) The effects experienced by any entity within the web of reality concerning a quantum system can only be predicted by the application of two seemingly arbitrary mathematical procedures.

Zurek's work rigorously explains both. The first, as might be expected, is the nub of the conundrum. Zurek shows that we should not expect the quantum world to be part of the objective web of existence that we inhabit; only the effects of quantum systems that can pass through the filter of decoherence compose our reality. Quantum systems participate in reality only through those interactions. When analysed in detail these interactions or quantum decoherences consist of a transfer of information between the quantum system and the web of objective reality. Not all information concerning the quantum system is transferable. In fact the vast majority is not transferable. The relatively tiny amount of information that can be transferred is selected from the huge range of quantum possibilities and numerous copies of this information are deposited in the environment by a Darwinian process which Zurek coins Quantum Darwinism.

Second Zurek has succeeded in showing that the two seemingly arbitrary mathematical procedures specified by the last two axioms required to predict the nature of quantum systems' interactions are inherent in the first three axioms. In other words once we concede that much of quantum systems exist outside of our web of objective reality, in a manner described by the first three axioms, the nature of their interactions within our reality follows.

In this manner Zurek's work leads us to a new tighter definition of objective ontological reality. It is the sum of the interactions between the fundamental entities composing the universe. Information or influence from anything outside this web that might, in some weak sense, be said 'to exist' cannot in principle be detected within this reality.

Over time this web of interactions, in spite of the 2nd law of thermodynamics, has found emergent paths to forms of greater complexity including atomic physics, chemistry, biology and culture. These forms are fundamentally composed of the objective web and are participants within objective reality.

The form of many of these emergent complex systems are well described in terms of the framework of an adaptive system as outlined in the work of Friston (2007) and others. Surprisingly, even the building blocks of our web of reality, as outlined by Zurek, also conform to this framework of adaptive systems. This suggests a possible line of reasoning for why quantum systems form the boundary of our web of reality:

1) The web of reality is composed of entities able to exchange information.
2) Quantum systems are the simplest information systems existing in nature.

3) All other information systems existing in nature are emergent from the quantum substrate.

Such a level of sophistication with quantum processes at the foundation of our web of reality indicates that much of the machinery inherent to adaptive systems may be a minimum level of complexity required to participate in the web of reality.

Computing the Future

A portentous event occurring in our time is the development of theory and technology concerning quantum computation. As implied by its name quantum computation is thoroughly quantum in nature. A number of quantum particles are entangled in an interactive web and their properties are manipulated to make their relationship logically analogous to a computational problem to be solved such as a prime number factorization. To extract the problem's answer a decoherence of the quantum web is forced and the system deposits its answer into the environment via Quantum Darwinism. In this way quantum computation is entirely equivalent to quantum interactions. As noted by Seth Lloyd of MIT, the universe can be seen as a quantum computer and each interaction between fundamental particles as a quantum computation. What is computed is the outcome of the interaction (Lloyd 2007). In other words what is computed is the future of the constituents composing the universe.

A formal equivalency (isomorphism) has been shown between the formulations of information transfer as presented by Quantum Darwinism and that presented by quantum computation (Blume-Kohout, et al. 2008). Given that quantum computation is analogous to quantum theory we must conclude that quantum measurements, quantum interactions and quantum computations all refer to the process of quantum decoherence occurring in differing contexts.

This observation justifies a merging of the interpretational explanation offered by Quantum Darwinism with that offered by Quantum Computation. Insight may be gained from hybrid interpretations such as: the future of the universe is selected by a Darwinian process. This interpretation suggests that the future history of the universe is selected from its potentialities through a Darwinian process, from possibilities generated in the past on the basis of their adaptability for survival or existence in their environment.

Given that the fundamental web of interactions at the quantum level may be interpreted as composed of a Darwinian process, it is little wonder that many of the best scientific theories describing scientific subject matter of greater emergent complexity (life, population genetics, neurology and culture) also rely on Darwinian processes.

Maximum Entropy

Accepting that a web of quantum interactions forms the basis of objective reality still leaves us with a puzzle as to the existence of complex structures, ourselves included. Complexity is an extremely special state and is therefore rare. This notion is a basic law of physics known as the second law of thermodynamics. It says that the quantity of entropy, or disorder in a closed system will tend to increase. That is because during the random walk entailed by the evolution of systems, all other things being equal, they are much more likely to enter states of lesser complexity than states of greater complexity due to the fact that less complex configurations are much more numerous.

How do we explain not only the existence of the many complex systems we see around us but also their increase in numbers and degree of complexity over time? The usual explanation is that entropy need only increase in a closed system and that the complex systems we are considering are not closed but rather part of a larger system whose overall entropy is increasing. For example, the complex biological system on earth is not a closed system but must be considered together with the sun. In this sun/earth system entropy is increasing since the decrease on the earth is more than compensated for by the increase of the sun.

Still this is only a partially satisfying explanation as it leads us only to more questions. How are designs for extremely unlikely complex systems such as earth's biology found? How are those designs instantiated, and how can they survive and evolve ever greater complexity?

A first step to answering these questions might involve a scientific principle that is a refinement of the second law of thermodynamics known as Maximum Entropy. This principle, although it may appear to only tweak the second law, offers enormous insight. It says that a system's entropy will increase except where constrained by scientific law from doing so.

As an illustration we might consider a historical scientific mystery concerning the atom. The atom was first envisioned as a classical entity with an electron orbiting a nucleus. According to classical scientific law the atom should achieve lesser complexity by radiating away its electrons' energy, the electrons subsequently spiralling inward and collapsing into the nucleus within a fraction of a second. This is a disastrous model as it predicts that all matter will be extremely unstable. Only when it was accepted as a scientific law that an electron's orbital energy came in quanta or packets and could not go to zero was it understood that an atom remained stable because its electron orbital energy was constrained from falling below its ground state. This quantum scientific law constrains the system's tendency to go to states of higher entropy and the situation conforms to the predictions of maximum entropy; the system will go to the states of highest entropy available to it subject to the constraints of scientific law.

Although some of the underlying principles of Maximum Entropy have been used by practicing scientists at least since the time of Laplace, it only became formalized during the last century largely in the work of E.T. Jaynes (1979). Since then it has become an indispensible implement in the scientific toolbox as it guides modelling of a system on the basis of the second law of thermodynamics and any other applicable scientific laws.

Maximum Entropy is usually considered as a principle providing a tractable method for making predictions on a system's evolution subject to scientific law but it may also be turned around and used to inform us as to the nature of scientific law.

Consider that in the very early universe scientific subject matter was quite limited. I am using the term 'scientific subject matter' in an objective sense. Scientific law greatly preceded humanity and has only recently been partially discovered by us. In the early universe the only scientific law in operation was probably that of quantum gravity. Atomic physics, cosmology, chemistry, biology and culture, in fact the bulk of scientific subject matter, had not yet emerged in the web of reality.

From this view it is clear that the bulk of scientific law came into existence along with the complex entities that could endure and that now after a long period of evolution compose much of scientific subject matter. Scientific law may be considered as the specific design details found by nature that are capable of retaining and evolving their complexity. Let us remember that complex systems are not forbidden, they are only extremely rare. A complex design able to maintain and evolve its complexity is not impossible, in fact we have overwhelming evidence that such systems exist. When we examine their nature we see that there are scientific laws operating which constrain these systems from evolving to more simple states. These scientific laws however did not pre-date the scientific subject matter composing the complex system; rather they evolved with it. The applicable scientific laws might be considered as their design specifications.

Thus we may consider scientific law governing such subjects as chemistry or genetics as the design specifications of those systems that nature has found and instantiated which are capable of retaining and evolving their complexity. They specify the constraints imposed by nature that deny the evolution of these designed complex systems access to states of lesser complexity.

Maximum Entropy is a useful principle in our quest to understand the nature of complex emergent systems as it leads us to consider the existence of complex systems as due to scientific law, developing over the course of evolutionary time, which serves to constrain the increase of entropy.

Darwinian Processes

It is clear that complex systems emergent from the web of reality owe their existence to the functioning of constraints which serve to maintain the system's low entropy state. These constraints have emerged along with their systems over the course of evolutionary time. A key step in understanding this process is to understand the manner in which such constraints are discovered and implemented. It is a central tenet of Universal Darwinism that the mechanism responsible is a Darwinian process.

The algorithmic nature of natural selection allows its essential mechanism to be abstracted and hypothesized as a possible mechanism operating in the evolution of both biological and non-biological realms. Numerous theories of this type abound in the social sciences and even in the hard sciences such as physics (J. Campbell 2009).

Darwin himself speculated on the idea that an algorithmic abstraction from natural selection might be useful in the explanation of the evolution of languages (Darwin 1872).

The essential abstraction from natural selection, which we are calling a Darwinian process, has been developed in the work of Richard Dawkins, Daniel Dennett and Susan Blackmore (amongst others) to consist of a three-step process:

1) Replication of information.

2) Inheritance of some characteristics that have variation amongst the offspring.

3) Selection or differential survival of the offspring according to which variable characteristics they possess.

It has been proposed that any system adhering to this three-step algorithm, regardless of its substrate, must evolve and will evolve in the direction of an increased ability to survive (Dawkins 1976) (Dennett 1995) (Blackmore 1999).

Indeed replication is a process central to the notion of a Darwinian process. Darwin saw this replication exemplified with parental organisms giving birth to offspring. However the discovery of DNA has since clarified the exact nature of this copying process; it is coded information in the form of DNA which is replicated and that forms the design specifications for the next generation. Thus natural selection is a process involving both a phenotype existing in ontological reality (what is really out there) as well as a genotype existing as an epistemological reality (a reality of knowledge or information).

As it has turned out the replication of information is a common feature of theories utilizing a Darwinian process. A model of some generality views the Darwinian process as involving:

1) An ontological parent; in the case of biology a phenotype.
2) An epistemic model internal to the parent; in the case of biology the parent's genotype.
3) A copied epistemic model with the design specifications of the child. In the case of biology the child's genotype.
4) A ontological child constructed from the copied epistemic model; in the case of biology the child phenotype.

The copied information typically forms the design specifications or model from which the new entity is constructed. Thus many of these theories explain the evolution of their subject matter as replication of an epistemic model composed of information followed by the forming of an ontological copy of the original (see table below).

This dichotomy between epistemic reality (to do with information or knowledge) and ontological reality (to do with what actually exists in reality), may be easily misunderstood. It is a theorem of information theory that information must have a physical representation. In other words, any epistemic substance must have a physical representation; that is, it must have an ontological existence. For instance the genome is the information of life but this information has an ontological representation in the form of DNA.

It is information which is copied in the replication step of a Darwinian process. We will examine a number of theories within Universal Darwinism in later chapters which describe the evolution of their subject matter in terms of this common process.

Theory	Epistemic form of copied information	Ontological form of copied information	Ontological result
Quantum Darwinism	Wave function	Outcome of quantum interactions or measurements	Quantum entities: atoms, molecules, etc.
Natural Selection	Genetics	DNA	Phenotypes
Population Genetics	Genetics	Distribution of alleles	Population of varied phenotypes.
Bayesian Brain & Synaptic Darwinism	Network of neural connections	Synapses	Adaptive behaviour
Memetics	Memes	Unknown mental constructs	Cultural constructs and artefacts.

Adaptive Systems

The understanding previously developed of the role played by Maximum Entropy and Darwinian processes in the creation and evolution of complex systems emergent in our web of reality may be brought together in a synthesis within the theory of adaptive systems.

Karl Friston (2007) models much of the subject matter of behavioural biology and neuroscience as an adaptive system involving three components: the state of the system, the effect of the system on the environment, and the effect of the environment on the system. Using this model he is able to arrive at a number of conclusions:

1) The system will possess internal models portraying external causes in the environment.
2) The system will evolve, through natural selection, to reduce the discrepancy between the internal models and the external environmental causes.
3) A generic scientific law can be proposed that predicts brain functions in numerous contexts; a minimization of the 'surprise', bounded by free energy, between internal models and external reality.

As Friston noted (Friston and Klass 2007):

> In summary, the free-energy principle can be motivated, quite simply, by noting that systems that minimise their free-energy respond to environmental changes adaptively. It follows that minimisation of free-energy may be a necessary, if not sufficient, characteristic of evolutionary successful systems. The attributes that ensure biological systems minimise their free-energy can be ascribed to selective pressure, operating at somatic (i.e. the life time of the organism) or

evolutionary timescales (Edelman 1993). These attributes include the functional form of the densities entailed by the system's architecture. Systems which fail to minimise free-energy will have sub-optimal representations or ineffective mechanisms for action and perception. These systems will not restrict themselves to specific domains of their milieu and may ultimately experience a phase-transition (e.g., death).

This finding may go some way to illuminating not only the nature of adaptive systems but more generally the evolution of emergent levels of matter within objective reality.

The brains of humans and other intelligent mammals are low entropy structures which require explanation. The principle of Maximum Entropy tells us to expect that their existence is dependent on constraints in the form of scientific law.

Evolutionary origins of brain functions can be traced to the abilities of early single celled life to perform phototaxis and chemotaxis. Such abilities clearly provide a selective advantage as they promote adaptive exchanges with the environment (Grabowsk, et al. 2008). Natural Selection, the Darwinian process involved, has selected progressive improvements in these abilities, ever more optimal mechanisms for adaptive exchanges with the environment. This progression is evident in examining such things as the evolution of neural chemistry and neural structures in the evolutionary chain of life forms from yeast cells to vertebrates (Emes, et al. 2008).

Natural Selection operates through the discovery and instantiation of adaptations such as sensing, perception and learning. Each of these adaptations has its own design details embodied in its functioning. The organisms possessing these adaptations are dependent on them for their survival. Superior adaptive designs are rare and, when found through the operation of a Darwinian process, tend to be adopted and copied in numerous guises. These powerful, widely-copied mechanisms must be rational; they must adhere logically to the objective nature of their environment. Essential details of a widely-replicated design, such as genetics within biology or optimal Free Energy structures in the brain, can sometimes be isolated and proposed as a scientific law or principle.

Successful designs arising in adaptive systems are those that can improve survivability through more optimal exchanges with the environment. Such designs specify lowered entropy states. The specific mechanisms discovered and selected by Darwinian processes to enhance survivability and lower entropy form the constraints that are responsible for preventing the system from moving to states of higher entropy. These designed constraints take the form of adaptations.

Friston's theory provides details of the operation of this process within behaviour and neuroscience, but the model it provides appears to be applicable to adaptive systems in general.

We might note a possible source of confusion between an adaptive system's internal model or design and the actual structures built from this design called adaptations. We might think of these two aspects of adaptive systems as the epistemic, or knowledge aspect, and ontological or physical aspect. However this view becomes somewhat muddied when we consider that epistemic aspects of the system must also have a physical representation. For instance the epistemic aspect of an organism is its genome but this has a physical representation in DNA. Perhaps a better way to view this is that the epistemic aspect contains the design from which the ontological structures, or adaptations, are built. It is the adaptations which serve to produce the reduction in thermodynamic entropy necessary for the complex adaptive system to exist.

3: Bayesian nature of the web of reality

As noted in the introduction, perhaps the most plausible explanation of mathematics' 'unreasonable effectiveness in the natural sciences' is that physical reality, the subject matter of science, is isomorphic to a mathematical structure. Science generally assumes this as a starting point. If some aspect of physically reality can be characterized mathematically then this characterization can be subjected to mathematical manipulations and analysis, the results of which may provide further insights into the physical process. The mathematical implications of the initial physical characterization are drawn out and these implications are assumed to be equally as valid as the initial characterization. The correctness of this assumption is dependent on the further assumption that nature is isomorphic to a mathematical structure.

The initial characterization may be given in the form of a small number of axioms. Such axioms contain all of the logical content of their implications. Thus the axiomatic treatment of quantum physics views all of quantum physics as implications of a few axioms.

Mathematics itself may be considered as an axiomatic system. Indeed since the beginning of the twentieth century it has been demonstrated in a number of ways that all of mathematics can be logically reduced to the implications of one group of axioms or another.

Often the axioms chosen as forming a basis for mathematics seem unintuitive and perhaps contrived. A surprising exception is the system developed by George Spencer-Brown (1979). In a stunning piece of mathematical virtuosity he demonstrated that the mere existence of a distinction or a boundary is sufficient. Using the usual definition of a distinction he notes that it implies two values, one for each side of the distinction, and he provides two logically-related axioms.

1) To cross a boundary again is the same as not having crossed it at all.
2) To call the value of a distinction again is the same as calling it once.

From here he proposes some symbols for a calculus of logic and using these derives Boolean algebra and thus mathematics. We are left to conclude that any possible physical reality containing distinctions or boundaries must be isomorphic to a mathematical structure of the richness implied by Boolean algebra.

It is interesting to note that advanced scientific theories for all the fundamental forces of nature which serve to demarcate the boundaries of our web of reality invoke the holographic principle (Smolin 2010) (Green 1999) (Verlinde 2010). This principle states that physical reality may be viewed, on the microscopic scale, as a collection of volumes demarcated by boundaries, and that the physics occurring inside each volume is isomorphic to information contained on the boundary itself.

Combining these ideas suggests that physical reality may have a rich mathematical structure simply because it involves boundaries or distinctions at its most fundamental level.

In any case scientific experience leads us to expect that all aspects of reality conform to mathematics. Knowledge structures found in reality should be no exception. The unique mathematical structure describing the construction of knowledge or inference is Bayes' theorem. We should thus expect that we will find a physical implementation of Bayes' theorem wherever knowledge is found in nature.

The web of reality is connected by quantum interactions which are well understood in terms of information exchange. Over evolutionary time more complex entities have emerged in this web, characterized by sophisticated design and organization. Information is defined scientifically in terms of probabilities. If we are to begin to understand the operation of this web we must first examine the role played by probability.

Bayesian probability, one of the two main schools of probability and the one we will focus on, considers probability as a measure of a state of knowledge (Wikipedia 2008). It extends Aristotelian logic, where variables are either true or false, to the logic of probability where variables may have continuous real values between 0 and 1 (false and true) and describes how it can be used to describe the process of inference, of accumulating knowledge. It has been proven, assuming only rationality and consistency, that Bayesian probability is the only mathematical system able to extend logic into probability.

As E.T. Jaynes remarked (1986) :

> *So, thanks to Cox, it was now a theorem that any set of rules for conducting inference, in which we represent degrees of plausibility by real numbers, is necessarily either equivalent to the Laplace-Jeffrey's rules, or inconsistent.*

Mathematics of information and knowledge

Bayesian probability is involved not only in the definition of information but also in the process of inference which is the extraction of knowledge from information. In this section I attempt to introduce the mathematics of this process.

As a note to those readers who tend to shy away from mathematics, the main purpose of this section is to show rigorous support for the arguments made concerning information and knowledge. A full appreciation of the formulas is not necessary. In most cases I attempt to restate the meaning of the formulas in words. Appendix A provides a concrete example of how this mathematics can be used.

Having provided this caveat I must say that the mathematics considered here are wondrous in that they rigorously describe the unique process by which information is processed and through which knowledge may emerge. In this regard they have great intrinsic beauty and may be worthy of the efforts required to understand them.

The holy grail of physics for the past several decades has been to understand all physical forces or interactions as a single phenomenon. This mission was the focus of Einstein's later research involving attempts at unifying the electro-magnetic and gravitational forces. As we have previously noted, given the unification of the microscopic forces into the standard model, many feel the 'Theory of Everything' is at hand in the form of a unification of the standard model with gravitation. Leading candidates for the final theory include String Theory and Loop Quantum Gravity.

Crucially both of the phenomena to be unified are understood in terms of interactions involving information exchange. Indeed, in many treatments information is a more fundamental concept than even space, time or matter.

Viewing information as central is not only an essential concept for the physical sciences but has also become an influential idea in biology and social science. Fortunately the mathematics for handling information is well developed within the field of Information Theory.

While information has become the fundamental concept of much of scientific subject matter it may not in itself be fundamental. Information theory defines information in terms of probabilities: Information (I) is measured in terms of the probability (P) of a particular outcome ($_n$) of a random event (ω).

$$I(\omega_n) = log\left(\frac{1}{P(\omega_n)}\right) = -log(P(\omega_n))$$

Essentially we receive information when we receive evidence concerning the outcome of a probabilistic event. The quantity of information received is a function (the negative log function) of the probability we assigned to the event's occurrence before receiving evidence of its outcome. If we considered the actual outcome as certain we gain no new information when it actually occurs; however if we considered the event highly unlikely then we receive much more information when it occurs. In this sense information measures surprise or the extent to which our experience varies from our expectations, or conversely, the amount of uncertainty that is removed when we learn the actual outcome.

Information is defined in terms of probability, which, in a Bayesian context, may be defined in turn as: "a measure of a state of knowledge" (Jaynes 2003). On first consideration information may seem an unlikely candidate as a fundamental explanatory entity, since its definition in terms of probability seems to suggest a very high degree of inherent complexity. This definition might suggest that probability and thereby information involves humans and their states of knowledge, which would necessarily involve extreme complexity.

First we should understand that this definition of probability must be considered in an objective and non-anthropomorphic light. Bayesian probability acknowledges that an observer's calculation of a probability is dependent on its prior information, which is subjective. However all observers with the same prior information should calculate the same probability and the more similar their prior information, the more similar their calculated probability. In this sense probability and information are objective.

'Observer' should also be taken in the scientific sense used by Einstein to refer to all entities capable of detecting an influence or force. These include highly inanimate entities such as test particles and are therefore, in general, non-anthropomorphic.

Still, this acknowledgement may not reduce information to a fundamental status in the sense of requiring no further explanation. Information is surprise and for an entity to be surprised may require a substantial amount of complexity. It requires a prior state of knowledge or some expectations of the outcome of a random event. It also requires that evidence of the actual outcome of the random event interacts with or is compared to this expectation.

While information is a measure of the surprise that we experience on knowing the outcome of an event, we might also be interested in evaluating beforehand how much information we should expect to receive on learning the outcome of an event. This involves the mathematical concept of 'expectation' which is defined as the sum of all possible outcomes weighted by the probability of each actually occurring.

For example if we are flipping a fair coin (P(heads)=.5, P(tails)=.5) and we will receive $1 for each head and $2 for each tail then our expected reward on each flip is .5*$1 + .5*$2 = $1.50. If we play this game for 10 turns the best estimate of our expected winnings is $15.

Expectation of information that will be gained on recording the outcome of a probabilistic event is called entropy. Entropy is calculated by summing the information of each possible outcome weighted by the probability for that outcome.[2]

$$E(\omega) = \sum_1^n p(\omega_n) \left(-\log \left(p(\omega_n) \right) \right)$$

Mathematically there are a few things we should note:

---

[2] The usual symbol used for entropy in mathematics and physics is H. Here I use the symbol E for entropy as H will be used for another entity, very important to our discussion: Hypothesis.

1) $\sum_1^n p(\omega_n) = 1$. This is a requirement that the sum over all possible outcomes must be 1. The probability that one of the outcomes will occur is certain.

2) Entropy is at its maximum when we assign a uniform probability to each of the possible outcomes and in this case has the value $\log_2(n)$. For example a fair dice with each of the 6 possible outcomes assigned a probability of 1/6 has greater entropy then a loaded dice where one outcome has a probability of ½ and the other 5 outcomes a probability of 1/10 each.

3) Entropy is a non-negative number which increases with the number of possible outcomes. For example a 24 sided fair dice has greater entropy than a 6 sided fair dice. In the extreme if there is only one possible outcome its probability must be 1 and log(1)=0 so the entropy of a certain outcome is zero.

Entropy may be thought of as the amount of uncertainty or surprise we expect will be removed when we learn the actual outcome.

If we have learned the outcomes of some events these may inform us as to the outcome of other events whose outcomes may be affected by those we already know. In other words knowing some outcomes may influence the probability we should assign to other outcomes and therefore the information we should expect to receive on learning them.

This brings us to the idea of mutual information. If we already know event Y on which the outcome of event X in some sense depends, then the additional information we should expect to receive when we learn the outcome of X is their mutual information:

$$I(X;Y) = E(X) - E(X|Y).$$

This may be read as: The mutual information of X and Y is equal to the entropy of X minus the Entropy of X given Y and may be thought of as the amount of uncertainty in knowing X that is removed when we know Y.

If an entity accumulates mutual information concerning other entities it might be said to have formed a model or hypothesis concerning those entities. For example the model of reality located in our brains contains mutual information with reality. This model serves to reduce the uncertainty we would otherwise have, for example as to which way an object will move when it is falling; our prior experience with other falling bodies leads us to expect them to fall down. As another example, an organism's genetics forms a model of its external environment and the mutual information contained in genetics serves to reduce the uncertainty an organism might experience concerning the types of resources available in its environment. It 'knows' what it should expect to encounter.

We might expect such hypotheses, being based on probabilities, to provide us with probabilistic predictions. We might also expect these probabilities to require updating as we receive new information. This process of updating the probability of a hypothesis being true upon the receipt of new information is called inference.

We will define knowledge as the characteristic of hypotheses or models to provide reduced surprise upon the receipt of new information. Thus knowledge is the ability of models to make accurate predictions of what will be encountered.

It should be noted that this definition of knowledge, as a reduction in expected surprise makes it practically synonymous with the definition of lowered entropy. This equivalency is useful as it leads us to expect that low entropy structures will contain knowledge mechanisms.

A more direct route to understanding the relationship between models and knowledge begins again with the definition of information.

$$I(\omega_n) = log\left(\frac{1}{P(\omega_n)}\right) = -log(P(\omega_n))$$

Information is a measure of the surprise associated with the outcome of some physical situation or variable. Its value can vary between zero and infinity. A precondition for information is the assignment of a probability for the outcome. This assignment must have a physical existence. The assignment of a probability to an outcome may be considered a hypothesis.

If probabilities are assigned to all possible outcomes of an event the value of these probabilities must sum to 1. This complete collection of hypothesis, covering all possible outcomes, may be considered a model of the event. Given that all possible outcomes are included within the model we may calculate the expected value of the surprise we will experience once the outcome is known. This property of the model is known as its entropy.

Entropy is the expected value of the surprise or of the expected inaccuracy of the model or the extent to which the model is expected to fail. We might be more interested with the accuracy of the model and the extent to which the model is expected to succeed. We have defined this measure to be knowledge or K and it is the negative of the model's entropy (K=-E).

Entropy is a maximum when the number of probability assignments is infinite and has a uniform distribution, meaning all of the probability assignments are equal. In this case the entropy is infinite and knowledge is at a minimum with a value of negative infinity.

Entropy reaches its minimum value of zero in instances where there is a single possible outcome with probability one. In this case knowledge is at its maximum of zero and the model is fully knowledgeable.

The second law of thermodynamics requires that the entropy in a closed system, such as the universe, must increase over time.

$$E_{u2} > E_{u1}$$

Where E_{u2} is the entropy of the universe at time 2 and E_{u1} is the entropy of the universe at time 1.

If the universe is divided into both a local system and the remainder of the universe outside of the local system we have:

$$E_{u2} = E_{u1} + Z_o + Z_l$$

Where Z_o is the increase in entropy of the universe outside of the local system between time 1 and time 2 and Z_l is the increase in the entropy of the local system between time 1 and time 2.

If entropy increases in both the local system and the universe outside of the local system between time 1 and time 2 both Z_o and Z_l are positive quantities and have the units of entropy. They are clearly entropy.

Entropy may be reduced in a local system only if the reduction in entropy is more than made up for by an increase in entropy in the universe outside of the local system.

In this case Z_1 is a negative quantity and the value of Z_0 must increase to compensate. Although Z_1 has the units of entropy, entropy cannot be a negative quantity. Z_1 does however fit the definition of knowledge.

This suggests that entropy cannot be reduced through entropy creation but may be reduced through the creation of knowledge.

How would such a knowledge mechanism function? Well clearly it must cause the model's hypothesis to become more accurate. In other words the model must learn to predict the behaviour of the event it is modelling more accurately.

Inference is the only mathematical mechanism known for this kind of knowledge creation and fortunately it is well understood. The unique mathematical process for accumulating or updating knowledge is Bayes' Theorem (Jaynes 2003).

$$P(H|IX) = P(H|X)\frac{P(I|HX)}{P(I|X)}$$

Although it may not be immediately obvious from this formula the key to accumulating knowledge is to have a model of an event consisting of numerous predictions or hypotheses concerning the possible outcomes of the event. Each hypothesis is assigned a probabilistic weight. As the model's hypotheses cover all possible outcomes of the event some will be more predictive of the actual outcome than others. When evidence of the actual outcome is received the probability of each hypothesis is updated in light of this new information using Bayes' Theorem. It is in this manner that the internal hypothesis or model may be kept in sync with those sources from which it receives information and honed to be more accurately predictive of future outcomes.

In the above formula capital P means the probability of the entity in brackets to the P's right. Thus Bayes' theorem has a probability on the left side of the equation and a function composed of three probabilities on the right hand side.

The capital H stands for the hypothesis being updated; it occurs once on the left and twice on the right and each of these instances signifies the same hypothesis, the one that is being updated. We must consider H as one of a family of competing hypotheses to which we have assigned probabilities of being true, that sum to 1. These hypotheses are mutually exclusive and one and only one must be true. Take as an example that a 5 will turn up on the next roll of a dice. The family of competitive hypotheses contains 6 members, that the numbers 1 through 6 will show up on the next roll of the dice. Each specific outcome is assigned a probability of 1/6. These probabilities form an expectation and we will be surprised and informed if the long term outcome does not meet our expectation.

There is one term, the denominator of the fraction that does not explicitly include H but it is still a factor even of this term. The term in the denominator is the probability we assign to receiving the particular new information given our prior information which includes a weighted average of probabilities over the competitive family of related hypothesis. This family includes H.

'I' stands for the new information we have received and are using to update our hypotheses.

'X' stands for our prior information or any evidence we may previously have received regarding this hypothesis. All four probabilities are given in terms of our prior information and this is an extremely important quantity. An example of the power of prior information in forming knowledge might be seen in regards to the simple axioms offered by George Spencer-Brown discussed in section 3. To the uninitiated (that is anyone without the right prior knowledge) these axioms may appear cryptic or perhaps only gibberish and as such are nearly meaningless to the recipient. The pertinent prior information consists of the ability to consciously model mathematical concepts and the ability to form expectations about them. With prior knowledge in place one might be able to follow Spencer-Brown's arguments and proofs, and through a chain of Bayesian updating, conclude that the hypothesis which states that these axioms imply mathematical structure is near certainty. Thus a difference in prior knowledge makes the difference between the plausibility of the hypothesis being updated to gibberish (near zero) or to near certainty (near 1).

Those of us having some but not sufficient mathematical background might be able to understand the significance of Spencer-Brown's argument and conclusions but not be able to directly verify the correctness of his proofs. In these cases we might search for critical reviews in the scientific literature of Spencer-Brown's work by qualified researchers. A peer reviewed conformation, such as the one by Kauffman (2001), might lead one to grant greater plausibility to Spencer-Brown's conclusions.

The vertical line in the formula means 'given' and is very helpful in translating the formula into something a little more understandable. For instance $P(I|X)$ is read as: the probability of receiving the new information given our prior information.

The wonder of Bayes' Theorem is that it describes exactly how expectations should be updated on the basis of new information in order to receive maximum knowledge. It is the mathematical mechanism by which information may be transformed into knowledge. Now we are in a position to outline the nature of the four probability terms used in this equation and to gain a better understanding of how knowledge may be increased.

The entity on the left hand side of the equation is the updated probability of the truth of our hypothesis: $P(H|IX)$

This is what we are calculating. It can be read as; the probability of our hypothesis being true 'given' everything we know; both our new information and our prior information.

The three probabilities on the right side of the equation are those we must use to calculate or update our knowledge:

$$P(H|X)\frac{P(I|HX)}{P(I|X)}$$

First let's consider $P(H|X)$.

This is the state of our knowledge prior to receiving the new information so it is the quantity that will be adjusted when we take our new information into account. This prior knowledge is updated via multiplication by a ratio of two other probabilities, the ratio $\frac{P(I|HX)}{P(I|X)}$

This ratio represents the change to our previous state of knowledge realized through gaining new information. It may be less than one or greater than one and is a function of our new information. If it is less than one the new information has made us less certain of the truth of our hypothesis; if it is greater than one it has made us more certain.

The numerator of this ratio: $\frac{P(I|HX)}{}$ is read: the probability of receiving the new information given that our hypothesis is true and given our prior information.

The denominator of the ratio $\frac{}{P(I|X)}$ is read: the probability of receiving the new information given our prior information.

The only difference between the two is that the numerator describes only the extent to which the new information supports the particular hypothesis we are considering. The denominator describes how well the new information supports a weighted average of all the hypotheses based on what we knew before receiving the new information. If the particular hypothesis being considered accurately predicted the new information better than the average hypothesis then the new information supports our hypothesis, the ratio of the two probabilities is greater than 1 and the probability of our hypothesis being true is increased over what it was before receiving the new information. If the hypothesis did not predict the new information or if it predicted information other than what was received the new information does not support the hypothesis and so our confidence it is reduced.

As a simple example we might consider Laplace's feat in using some new astronomical data in conjunction with Bayesian updating to arrive at an extremely accurate estimate of the mass of Saturn. While both the prior information and the new information used in this example are historically inaccurate they do illustrate the method.

Prior to Laplace taking on the project of estimating Saturn's mass this quantity was known to some degree of accuracy due to the fact that Saturn had sufficient mass to retain its moons in orbits but not sufficient mass to perturb the orbits of nearby planets more than was observed.

This prior information could be used to construct a model for the mass of Saturn composed of three hypotheses:

H_1) Mass of Saturn is between .000265 and .000280 solar masses. $P(H_1) = .2$

H_2) Mass of Saturn is between .000280 and .000290 solar masses. $P(H_2) = .6$

H_3) Mass of Saturn is between .000290 and .000300 solar masses. $P(H_3) = .2$

We can calculate the entropy, or expected uncertainty, of this model:

$$E_1 = \sum_{n=1}^{3} -p_n \log (P_n)$$

$$= -.2 * \log_2(.2) - .6 * \log_2 (.6) - .2 * \log_2 (.2)$$

$$= 1.370951 \text{ bits}$$

The new data indicates a value for Saturn's mass of 0.000285 solar masses with an experimental accuracy of +/- .000005 solar masses 99 times in 100.

In calculating $P(H_1|XI)$ we take $P(H_1|X) = .2$ as given above, $P(I|XH_1) = .005$ as the experimental accuracy allows the measurement to be in this range 5 times in a hundred and $P(I|X) = .2 * .005 + .6 * .99 + .2 * .005$, the weighted average of the probability distribution reflecting of our prior information.

Using Bayes' theorem new probabilities can be calculated for the three hypotheses composing the model:

$$P(H_1|XI) = P(H_1|X)\frac{P(I|XH_1)}{P(I|X)}$$

$$= .2 \left(\frac{.005}{.2(.005)+.6(.99)+.2(.005)}\right)$$

$$= 0.001678$$

Given the symmetry of the problem we see that $P(H_3|XI) = P(H_1|XI)$ so $P(H_3|XI)$ is also equal to 0.001678.

The results of these new accurate measurements provide increased confidence in our 2nd hypothesis:

$$P(H_2|XI) = P(H_2|X)\frac{P(I|XH_2)}{P(I|X)}$$

$$= .6 \left(\frac{.99}{.2(.005)+.6(.99)+.2(.005)}\right)$$

$$= 0.996644$$

On the basis of the new data the probability distribution of our hypothesis has become much more peaked around H_2, that is we have greatly increased confidence in H_2 being true. As the updated distribution has moved away from a uniformed distribution and as the certainty of our model has increased we should expect its entropy will have decreased. We can calculate the entropy of this updated model:

$$E_2 = \sum_{n=1}^{3} -p_n \log (P_n)$$

$$= -0.001678 * \log_2(0.001678) - 0.996644$$

$$* \log_2 (0.996644) - 0.001678 * \log_2(0.001678)$$

$$= .0.035773 \text{ bits}$$

As our ability to accurately predict Saturn's mass has increased, the entropy of our model has decreased to near zero. In terms of knowledge we can say that the knowledge of our model has increased by 1.370951 - 0.035773 = 1.335178 bits.

As entropy may never be reduced below zero we can conclude that there is little room left for increasing the knowledge of the updated model. Indeed during the intervening two hundred years with ever refined measurement techniques and Newton's theory of gravitation being replaced by Einstein's, the current best estimate of Saturn's mass differs from that of Laplace's by less than 1%.

The accuracy of Laplace's result hints at the power of inference in general and science in particular. Even though science evolves and the best models or theories are sometimes replaced by even better ones, those conclusions of the older models which have been confirmed by experimental data are refined rather than overthrown.

I will argue in the following sections of this book that the power of inference is employed, not only in the evolution of science towards greater knowledge, but in the evolution of all knowledge entities.

Bayes' theorem may be processed in an iterative manner, with the probability of the hypothesis calculated at one step used as the probability which is updated in the subsequent step, when the next piece of new evidence is received. It is in this iterative manner that hypotheses may evolve ever greater accuracy over time.

Innovations with the mathematical and computational techniques that describe complex real world Bayesian models have exploded in recent decades. It turns out that many of the techniques required for performing Bayesian analysis are very well suited to computer calculation. Besides the iterative application, other methods include the generalization to continuous probability distributions, the description of models in terms of parameters which may have a range of values, and the use of methods such as Markov chain Monte Carlo for randomly sampling a model's parameter space.

The mathematical procedure described by Bayes' theorem may seem simple, perhaps almost trivial and unworthy of being the source of all knowledge., However nature has learned to use this process in numerous settings in a complex and intricate manner. Often the internal models of adaptive systems are seen to be composed of a vast array of competing hypothesis and models, each receiving feedback through its experience in the world and updating itself accordingly. Mathematics tells us that an implementation of Bayes' theorem is the essential mechanism which must operate in the process of accumulating knowledge.

Perhaps we should not expect physical implementations of Bayes' theorem to be obvious. Physical implementations of knowledge mechanisms face numerous design constraints in addition to that of conforming to Bayesian mathematics.

Still we are guided by an expectation that any knowledge mechanism implemented in the physical world will share as an essential characteristic the Bayesian updating of probabilistic models.

4: Physical implementation of Bayesian processes

From mathematics to the physical world

We assume that entities in physical reality must operate in accordance with the structure of mathematics but they also must conform to another crucial constraint. The second law of thermodynamics and its extension via the fluctuation theorem, perhaps our most fundamental physical law, instruct us that entropy always tends to increase and that this tendency is exponentially pronounced with both time and increasing mass.

It turns out that these two constraints, mathematics and the second law, severely limit the nature of physical entities. If we define physical entities to be material systems having a particular structure or identity capable of existing over time, then the deck is stacked against the existence of any entity. The second law demands that entropy will increase thus washing away any particular structure or identity.

The prospects for survival is even more challenging for entities having greater complexity. Because they have greater mass they are doomed by the fluctuation theorem as being even far less likely to persist with a specific structure or identity.

There is, however, a particular loophole in this theorem that nature has found and exploited. The fluctuation theorem applies to closed systems, those having no energy exchange with systems outside themselves, leaving open the possibility that entities that exchange energy with other entities might be able to avoid an increase in entropy, as long as entropy increase is the net result over all entities involved.

Although it is possible for an open system to avoid an increase to its entropy it is extremely unlikely to occur accidently. It is only possible to maintain a complex entity in a low entropy state through the use of knowledge.

Entropy increase is generally vast as it enumerates all the potential states through which a system could evolve. Even a small decrease in entropy experienced by a simple system over a short period of time is extremely improbable; it doesn't happen by accident. A decrease in an entity's entropy requires that the energy flow between it and other systems is of an extremely rare and specific nature; that is the energy flow must be designed to maintain the particular structure or identity of the entity under consideration. This design must encompass knowledge of other systems in the environment with which the entity may have interactions and this knowledge must model a shifting of entropy from the entity to its environment in a manner that realistically allows the entity to survive intact. Further, the expectations or predictions of the model must be accurate; they must actually occur.

We have defined model accuracy as knowledge and mathematics tells us that such knowledge can only be created through inference or Bayesian updating. We can conclude that a common characteristic of physical entities is their capacity to build knowledge.

Thus we find that knowledge, based on information, is necessary for an entity's physical existence. Mathematics requires that this knowledge takes the form of a model whose accuracy evolves through a physical implementation of Bayesian updating.

Accurate knowledge of an effective design while necessary is not by itself sufficient for the existence of entities. This knowledge must be physically instantiated; it must result in the construction of those physical objects specified by the design. These physical objects will often form constraints that restrict the increase of entropy in conformance with the principle of maximum entropy.

Both the knowledge of the design and the knowledge required for the construction of physical objects specified by the designs must be contained within the internal model. That is, the internal model must contain not only design knowledge but also construction knowledge.

Once the design is constructed in the physical world it is tested there. Does the design accurately and realistically envision the entitiy's continued survival? Does the construction knowledge contained in the model accurately and effectively implement the design? Survival of the entity is dependent on the constructed design accurately reflecting the model.

Failure on these tests may cost the entity its existence but a further hurdle is presented to those that pass. Mathematics requires that the entity's model must be updated with the test results if it is to retain knowledge of a changing world. There must be a mechanism for incorporating evidence of the more effective designs and construction methods back into the model so as to maintain its accuracy; that is, inference must be performed.

In this light even the most rudimentary physical entities are information based and have surprising complexity. They must have:

1) Design knowledge specifying an entity able to survive and the knowledge required for constructing this design in physical reality.
2) An internal model having a physical representation containing this knowledge.
3) An ability to construct the design in the physical world.
4) An ability to register evidence of the success of the constructed design. These are real world tests which result in differential survival of different designs.
5) An ability to update the internal model with this evidence.

These five functions describe a cycle which all physical entities must perform. In order for an entity to have extended existence its design must keep current with a changing environment. This requirement may be realized by performing the five step cycle iteratively where #5 of one cycle describes a model containing the knowledge referred to in #1 of the next cycle.

The model of step 5 is a copy, with perhaps some variation, of the model of step 1. Steps 4 and 5 describe differential survival. Thus our cycle closely approximates the Darwinian process.

We have argued that the existence of complex entities demands a specific interplay between information and physical reality that is analogous to the Darwinian process. I suggest that it is in these terms that we should understand Universal Darwinism as a scientific theory which attempts to explain the creation and evolution of those entities composing scientific subject matter in terms of the Darwinian process.

Darwinian processes, Bayesian updating and adaptive systems.

It is perhaps understandable that science has been slow in coming to the fundamental understanding that any complexity is extremely unlikely and deserves explanation. After all, our universe is populated with complex objects that compose, almost exclusively, the content of scientific subject matter. It is easy to take them for granted.

When science does take this conundrum seriously it can offer only a single explanation: the Darwinian process. Lee Smolin, a prominent physicist and founder of the theory of Cosmological Natural Selection makes this point (2005):

> There is only one mode of explanation I know of, developed by science, to explain why a system has parameters that lead to much more complexity than typical values of those parameters. This is natural selection.

Darwin himself understood that the essential mechanism of Natural Selection could be abstracted and used to comprehend the nature of evolution operating in non-biological processes. This powerful idea of a Darwinian process was largely ignored, outside of biology, until the twentieth century when it was again championed by the work of Richard Dawkins (1976). A short chapter of his book The Selfish Gene introduced the theory of *memetics* as a possible Darwinian explanation of cultural evolution. This concept was further developed in the work of Susan Blackmore and Daniel Dennett, amongst others (Blackmore 1999) (Dennett 1995).

However the 'copy with selective retention' algorithm of a Darwinian process is somewhat ambiguous. What is it that is copied? Darwin assumed, correctly on the basis of the evidence available to him, that it is the phenotype that is copied between generations in the process of natural selection. Since the discovery of DNA we know that it is not quite so simple. The copying process is actually done in two steps. First the information forming the entity's model (genome) is copied then the phenotype is constructed from this copied model forming a copy of the original phenotype.

The absence of DNA from Darwin's original formulation of natural selection has led to some confusion in its generalization to definitions of the Darwinian process. Many definitions such as 'copy with selective retention', fail to make clear what is copied. Perhaps some researchers have concluded that the simple algorithm works as long as it is carried out and that the exact nature of the copying process in unimportant to the big picture.

However an understanding that natural selection - and more generally - Darwinian processes perform their copying in two steps means we can integrate Darwinian processes seamlessly with information theory and the theory of adaptive systems. It is useful to understand that the copying process first makes a copy of an internal model, or epistemic component, from which a copy of the original entity, or ontological component, is constructed.

I have argued on theoretical grounds that a generalization of this duality is a unique solution for constructing complex entities within physical reality. The essential reason for this is that complexity requires the exploitation of loopholes within the second law and this requires a model containing knowledge of the environmental opportunities. In order for the entity to exist in physical reality it must be constructed from this epistemic component of knowledge.

Dawkins and Blackmore take a complementary view on this issue. They argue that a useful copying or replication mechanism within a Darwinian process must have high fidelity; it must be highly accurate. This is because any entity able to exist must already possess an excellent design. Any random alteration is likely to be harmful, though some variation must be possible to provide potential for improved copies. Therefore the accuracy of the copying process must be kept in bounds and those bounds call for a high degree of fidelity.

One general method proposed for maintaining high standards of fidelity within Darwinian processes is that the copied entity be in a digital format.

This brings us to a bigger question? How, in principle, would it be possible to make an essentially identical copy without invoking information? In cultural affairs things may sometimes be copied through reverse engineering the original, but this still entails inferring a model for the original and then constructing a copy from it. More commonly we copy the instructions. Indeed this is essential to much of learning; memorize (copy) the instructions so that you can then carry them out. In general we make copies by following the process (which might be condensed) by which the original was made.

Instructions are a form of information or knowledge and may be easily copied, especially in a digital format. Of course once the necessary information is copied there must be a mechanism capable of constructing the entity using the information much in the way a cook constructs a dish using a recipe.

However information that specifies the original is not enough; an actual copy or construction must be made from the information in physical reality. The construction process is also specified in the model as are the expected resources available in the environment that are required by the construction. Failure will be the result if these expectations are not met.

Dawkins' playful assertion that organisms are only the gene's method of making more genes seems to turn common sense on it head but we might question why do the genes need an organism, or in a general sense why does the epistemic require the ontological. The answer is: the epistemic must have a physical representation; the internal model must have a physical existence and be subject to the second law. We might understand the organism, or more generally the ontological aspect of an entity, as a constructions undertaken by the epistemic to shield and immunize itself from the second law.

Within the copying process it is possible that the copy is being constructed in a slightly different environment than the original, and that one of the crucial resources is in less abundance. This is where variation plays its role. If there are a sufficient number of copies, each with the potential for slight variation, then at least one is likely to have a variation calling for a little less of that particular resource.

We can push the recipe analogy a little further. In order to prepare a successful dish the supplies and equipment specified by the recipe must be available to the cook. If the resources specified in the recipe are not adequately available to the cook the dish will be a failure. A solution is to find a variation of the recipe still describing a great dish but using only the resources available.

Lastly a Darwinian process involves selection. The general rule is that those entities able to pass their real world tests have more chance of a continued existence than those that do not. This process leaves physical reality populated with selected individuals. Mathematically this result is a Bayesian update. The models belonging to the surviving population have been updated through experience and tend to contain greater accuracy than those of the previous generation. That is, they contain more realistic expectations.

The Darwinian process may be the only physical mechanism known to science capable of accumulating knowledge from experience. It performs inference and is a physical analogue of iterative Bayesian updating. This analogy may be stated in terms of the 3 step Darwinian algorithm:

1) Replication. Iterative Bayesian updating requires that the updated hypothesis of one cycle is copied into the next cycle where it is in turn updated.
2) Variation. There must be some variations in the copied hypothesis. In a changing world any model of it must also change if it is to retain accuracy. Variations in the copied hypothesis may include new ones that more accurately reflect features of the world and some with less accuracy. This range of hypothesis provides a model which may be updated through Bayesian inference.
3) Selection. The awarding of greater probability to hypotheses having greater accuracy and the awarding of lesser probability to those having less accuracy is achieved by the Darwinian process. Differential

survival of offspring ensures that the fittest survive. Fitness should be understood as accurate expectations concerning the entity's interaction with its environment. The distribution of variable characteristics or hypotheses will shift from generation to generation in a manner enhancing survival or accuracy.

Another area of research providing insight into complex entities is that of adaptive systems. Karl Friston (2007) defines an adaptive system as ones that can react to changes in its environment by changing its interactions with the environment to optimize the results for itself. He examines the system in term of three features or variables: the system itself, the effect of the environment on the system and the effect of the system on the environment. These three features are integrated via an internal model that resides within the system. Friston has shown that the measures taken by adaptive systems to optimize their interaction with the environment can be understood via the Free Energy Principle: an adaptive system will attempt to minimize the free energy which bounds the 'surprise' resulting from its actual experience in the environment. In other words the system will act so as to minimize the prediction error, and thus maximize the knowledge, of its internal model.

The internal model is constructed in a Bayesian process that makes inferences from environmental data and also utilizes knowledge of the actions which the system can take to influence outcomes in the environment. Examples of such internal models residing within adaptive systems may include genetics residing within organisms, mental models residing within brains and business plans residing within corporations.

Humans are adaptive systems and continuously conform to the paradigm although we might easily miss this fact. Let's say we see something out of the corner of our eye that might be important. Because we did not see the object very clearly we may have trouble assigning accurate probabilities to our hypothesis space of the numerous imaginable possibilities. Being adaptive systems we might turn our head and/or eyeballs and bring the mysterious thing into focus. We capture a pertinent data set and update, in a Bayesian manner, the probabilities associated with our internal model of the external world in conformance with this data. One particular possibility achieves near certainty at the expense of the others. We change the way we interact with our environment by bringing a potentially important aspect of our environment into focus. We are now better informed and in a better position to optimize our response.

The system's interaction with its environment must be designed to optimize the outcome for itself. Thus an adaptive system must envision an optimal but realistically possible outcome of environmental interactions and execute those actions under its control necessary to achieve the envisioned outcome.

From this view the modeling performed by an adaptive system functions as a simulation of its environmental interactions and is thus a form of computation. The theory of computation as outlined by Turing and the Church-Turing-Deutsch principle makes the claim that there exist universal computing machines capable of processing all computable functions. David Deutsch (1985) has shown that the composition of this underlying class of computable functions is not defined by logic or mathematics but rather by the laws of nature. What is computable are those functions where nature has provided the machinery capable of performing the algorithm. In Deutsch's words:

> The reason why we find it possible to construct, say, electronic calculators, and indeed why we can perform mental arithmetic, cannot be found in mathematics or logic. The reason is that the laws of physics 'happen to' permit the existence of physical models for the operations of arithmetic such as addition, subtraction and multiplication.

Some researchers consider the computational aspect of adaptive systems as central to understanding them. Nobel laureate Sydney Brenner (1999) has called for biological science to focus on developing the theoretical capabilities required to compute an organism's phenotype from its genotype. This approach is judged to have explanatory potential as he characterizes biological systems themselves as essentially information-processing machines.

It is interesting that we find computational machinery central to the functioning of adaptive systems and that it is such computational machinery found in nature that underlies the theory of computation.

An adaptive system is defined in terms of the effects the system can have on the environment and the effects the environment can have on the system. These are effects which occur in the real or ontological world and do not directly involve the system's internal model. Any mediation that adaptive systems may exercise on these interactions is achieved through adaptations the system has constructed. The physical entity may often be understood as composed almost entirely of numerous adaptations. In this light we should understand adaptations as constructions designed by the internal model which mediate the entity's interaction with its environment. The design goals of these constructions are to protect the entity from states of lowered entropy and to enhance the likelihood of survival.

Adaptive systems, according to our discussion, contain a good deal of internal machinery. They are complex, low entropy structures. The primary outcome of environmental interactions that any adaptive system must achieve in order for it to be an optimal encounter is to retain its low entropy structure, to survive intact. This is not a trivial accomplishment.

An adaptive system must orchestrate encounters with its environment in such a way as to maintain its own low entropy state at the expense of increased entropy in the environment. Selecting and executing these rare outcomes requires knowledge; both knowledge of the causes operating in the environment and knowledge of how other entities existing in the environment can be exploited to maintain the system's low entropy state.

For example if we consider life forms as adaptive systems we should expect them to utilize adaptive mechanisms for lowering entropy. Sunlight is an abundant energy source at the earth's surface but only extremely specific encounters with this energy will result in work being done in the service of local entropy reduction. Life acquired knowledge to perform this feat billions of years ago when it evolved photosynthesis and this knowledge has since become ubiquitous throughout the plant kingdom. Photosynthesis causes energy in sunlight to be transferred to a high energy bond in molecules of ATP. The energy in this molecular bond is then used as a resource for performing work through complex chemical processes which in the end serve to retain the organism's low entropy state.

The knowledge required to perform photosynthesis and to use the ATP molecule is contained in the genetic code of organisms and this knowledge is passed between generations through copying of the organisms' genomes. These adaptations were created and have evolved to their present states due to the operation of the Darwinian process called natural selection.

Systems must move to states of highest entropy consistent with the constraints operating on the system. The ability of adaptive systems to maintain their low entropy state must therefore be explainable in terms of the constraints they deploy against increasing entropy. These constraints are precisely the adaptations developed during the evolution of the system. In our example of the production and consumption of ATP molecules those constraints often take the form of enzymes which constrain the system's chemical pathways to a specific course. These enzymes are directly coded for in the organism's DNA. Thus the knowledge implicit with their entropy-reducing abilities is contained in the organisms' internal models.

At the heart of Bayesian probability is the notion of updating: our belief in a hypothesis should be updated whenever pertinent new data becomes available. Bayes' Theorem provides the exact measure by which our confidence in a hypothesis should change with a given piece of new data. The essence of the theorem is the common sense notion that our confidence should be adjusted in accordance with the extent to which the hypothesis or model enhances our ability to predict the new data.

This theorem has been shown to be the unique mathematical description of how knowledge may be accumulated (Jaynes, 1985). It is a signature characteristic of a Darwinian process and is a central mechanism in an adaptive system's ability to model its situation (J. Campbell 2009). A consequence of the importance of Bayesian updating is that we see evidence of its operation wherever knowledge accumulates in nature, in areas as diverse as genetics, brain functioning and the cultural process of science.

In the course of this analysis we have drawn on the theory of Bayesian processes and adaptive systems to provide additional insights into the nature of Darwinian processes. Given these insights we might construct a list of criteria we should expect to find in scientific theories utilizing the Darwinian process:

1) The operation of a three step 'copy with selective retention' algorithm.
2) Operation in terms of an adaptive system where the system, the effect of the environment on the system and the effect of the system on the environment are modeled internally and conform to Friston's free energy principle. The simulation of the system's environmental interactions may be understood as computation.
3) Maintenance of the internal model's accuracy through Bayesian updating.
4) Construction of adaptations, in accordance with the design specifications of the internal model. The adaptations provide constraints to the system's entropy increase as well as enhanced survivability to the system. These constraints may be understood in terms of the principle of maximum entropy.

Criteria 1 may be somewhat repetitious given criteria 3 as we have argued that the Darwinian process is a physical analogue of iterative Bayesian updating but this may not be a consensus view. In some theories such as the Bayesian brain the Bayesian criteria is well developed while the nature of the Darwinian process which carries it out is unspecified. I will retain these two as separate criteria and will attempt to explain their relationship in the theories examined in later chapters, as this relationship may not always be obvious.

Criteria 2 and 3, with their focus on models and updating emerge directly from our analysis of a web of reality characterized by information exchange. Criteria 4 emerges from reasoning that any physical implementation of a Darwinian process must be complex and thus must contain a mechanism for creating a lowered local state of entropy. In arguing that criteria 1 is but a physical implementation of criteria 3, I would suggest that criteria 2 through 4 provide a complete set of criteria for theories within Universal Darwinism that are directly derivable from information theory and fundamental physics, and that these criteria serve to place Universal Darwinism on a firm theoretical foundation.

The argument I am making is based on the self-referential nature of information. On the one hand information implies complex machinery including internal models, their comparison to actual experience, etc. On the other hand complex machinery implies information or knowledge required to circumvent the second law of thermodynamics. Ever since information-based-entities appeared within our web of reality there has been an ever escalating race to gain knowledge in order to support complexity and to gain complexity in order to support knowledge.

There is certainly no scientific consensus that this level of complexity must be a common property of even the most fundamental entity. Aren't we on the verge of the 'final theory of everything' which will explain all? I think not. While the explanation of much of scientific subject matter within a single simple paradigm may be within our grasp this step forward may only open new frontiers and still leave much unexplained. I expect that when science is at last able to gain an understanding of phenomena occurring within the twenty orders of magnitude between the Planck scale and the scale of 'fundamental' particles much will be added to our scientific understanding.

Even the particles of particle physics, the simplest entities currently examined by science, must encompass the complexity inherent in our list of criteria. These are far from simple and a theory uniting all particles and forces will not be the end; there will still be much to explain. It appears that even the most fundamental entities known to science are emergent from a deeper level of reality of which we are now largely ignorant.

Fortunately for my argument there is a powerful formulation of quantum physics, within the scientific literature, which largely conforms to our list of criteria. Hopefully this breakthrough may provide a path forward.

I have constructed our list of criteria in a theoretical manner by arguing that any complex entity must adhere to them. Now we are in a position to examine some of the most ambitious theories within Universal Darwinism, scientific theories which attempt to explain a wide range of complex entities and to discover the sense in which those theories meet our listed criteria.

5: Atomic physics and chemistry

As soon after the big bang as allowed by cooling temperatures, the light elements formed according to the predictions provided by quantum mechanics. Indeed the success of quantum theory in predicting the observed ratios of the light elements is considered some of the strongest evidence for modern theories of cosmology. The further development of more complex elements within stars and supernovas is also fully explained by the operation of quantum theory within these environments. Further evolution of more complex chemistry has been possible in more specialized environments throughout the universe (Dishoeck 1998). In particular the earth with its moderate temperatures, sources of solar and tectonic energy, and liquid water has proved an ideal environment for the production of complex chemistry.

Quantum theory has enjoyed great success in explaining all of these evolutionary advances of complex chemistry. The interpretation emerging from Zurek's quantum theory (2004) is that the states of matter composing atomic physics and chemistry are the results of quantum decoherence that occurs when the underlying quantum states interact with their environments.

Most of atomic physics and chemistry are derivable, at least in principle, from quantum theory with the assistance of other basic physical laws such as thermodynamics. Few emergent laws are required to explain the scientific subject matter of atomic physics and chemistry although there may be some exceptions in the realm of complex chemistry (Prigogine 1984).

Quantum Darwinism is a recent theory, well documented in the scientific literature and gaining experimental support. It makes clear many of the mysteries which have plagued quantum theory for the last century. As its name suggests it describes quantum theory in terms of a Darwinian process and thus opens to analysis the evolution of matter from fundamental particles to complex chemistry in terms of our list of criteria.

Quantum Darwinism

The research program into the nature of quantum systems conducted by Wojciech Zurek of Los Alamos National Laboratory and colleagues has spanned more than 25 years and has identified and developed a number of novel processes and constructs. The most important of these might include: decoherence, einselection, envariance and the theory of Quantum Darwinism. The theoretical advances gained by this program have enabled Zurek to derive what have commonly been considered two axioms of quantum theory from the remaining three.

Quantum theory is often presented as a derivation from a set of axioms. A common set of axioms are:

1) The state of a quantum system is represented by a vector in its Hilbert space.
2) Evolutions are unitary (i.e., generated by Schrodinger equation).
3) Immediate repetition of a measurement yields the same outcome.
4) Measurement outcome is one of the orthonormal states, the eigenstates of the measured observable.
5) The probability p_k of finding an outcome $|s_k\rangle$ in a measurement of a quantum system that was previously prepared in the state $|\psi\rangle$ is given by $|\langle s_k|\psi\rangle|^2$.

Decoherence occurs when information is copied from a quantum system to its environment. This process, often referred to as a measurement, has proven an unresolved interpretational quandary for quantum theory since the Bohr-Einstein debates that occurred during the first part of the last century and has become known as the *measurement problem*. The problem arises as many consider the axioms to contain seeming contradictions. Axiom 2 requires a quantum system to evolve via the Schrodinger equation in a mathematically smooth, deterministic and continuous manner, whereas axiom 4 and axiom 5 require the quantum system to jump in a discontinuous manner, upon measurement, to a specific state, one often much different from the state to which it had evolved prior to the measurement.

Zurek's resolution (2009) is to show that the two problem axioms are implied by the other three. The proof of this derivation reveals that much of the information contained in the state vector after a period of unitary evolution cannot survive the process of being copied into the environment. Only a subset can survive the transfer. Axiom 3 requires that the information that survives in the systems state vector after the transfer, is consistent with that copied to the environment, resulting in a seemingly discontinuous change in the state vector.

Theoretically, this understanding of decoherence views it as two processes roughly equivalent to axioms 4 and 5 (Zurek 2009). The first process is named 'environment induced superselection' or einselection. When a quantum system becomes entangled with an entity in its environment, properties of the environmental entity determine the type of information which can survive being copied from the quantum system into that environment.

For instance some environments may be structured to receive information on position and others to receive information on momentum. Mathematically the subset of information that can survive transfer to an environment is restricted to pointer states of the system. These pointer states are the eigenstates of axiom 4 and may be predicted as they are those which will leave the system in the lowest entropy state available (Zurek 2009). Conversely the ability of the environment to receive information from the quantum system is dependent on its ability to absorb this entropy increase (Zwolak, Quan and Zurek 2009).

The second major simplification of quantum theory produced by Zurek's program was the derivation of axiom 5 using symmetry properties of the entangled quantum system and environmental entities. This process, 'environment–assisted invariance' or envariance, provides a probabilistic prediction of the measurement value.

Importantly both of these simplifications provide predictions regarding measurement; they predict the type of information that will be measured and the value of the measurement. Given Zurek's demonstration that, in quantum processes, information is physical and that there is no information without representation and that axiom 1 posits that the state of the quantum system is represented by the Hilbert state vector, we are compelled to conclude that the quantum system must contain a physical representation of its state vector. Crucially, this physical representation may be considered an internal model residing within the quantum system, and this model, through the measurement predictions it contains, simulates the quantum system's interactions with its environment.

Recently Zurek's research has focused on the fate of quantum information that has been copied into its environment. His major finding is that the information best able to survive the process of decoherence is also the information having the highest reproductive success in the environment, thereby causing it to become the most widespread. This principle, along with decoherence, is encapsulated by the theory of Quantum Darwinism.

In Section 4 we reviewed suggested characteristics of Darwinian processes operating in physical reality. They are:

1) The operation of a three step 'copy with selective retention' algorithm.
2) Operation in terms of an adaptive system where the system, the effect of the environment on the system and the effect of the system on the environment are modeled internally and conform to Friston's free energy principle. The simulation of the system's environmental interactions may be understood as computation.
3) Maintenance of the internal model's accuracy through Bayesian updating.
4) Construction of adaptations, in accordance with the design specifications of the internal model. The adaptations provide constraints to the system's entropy increase as well as enhanced survivability to the system. These constraints may be understood in terms of the principle of maximum entropy.

Quantum Darwinism as a Darwinian process

Although Zurek's theory has 'Darwinism' in its name it has not been made clear in the literature in what precise sense his theory might be considered Darwinian. In order to explore this subject I will attempt a review of the manner in which Quantum Darwinism might be understood to embody the principles of Darwinian processes listed above.

Clearly, in Zurek's view, there is little question of the Darwinian nature of quantum processes or of its central importance; he sees Quantum Darwinism as conforming to the Darwinian paradigm and identifies it as the mechanism responsible for the emergence of classical reality from the quantum substrate (2003):

> The aim of this paper is to show that the emergence of the classical reality can be viewed as a result of the emergence of the preferred states from within the quantum substrate thorough the Darwinian paradigm, once the survival of the fittest quantum states and selective proliferation of the information about them are properly taken into account.

The Darwinian paradigm, as defined within Universal Darwinism, encompasses details in addition to survival of the fittest and selective proliferation. The following evaluation will consist of a comparison of the four listed characteristics of Darwinian processes with the characteristics of quantum systems as described by Quantum Darwinism.

1) The operation of a three step 'copy with selective retention' algorithm.

Decoherence essentially describes a process by which information is copied or transferred from a quantum system to its environment. Much of the varied information contained in the state vector is copied but most has extremely short periods of survival. The fittest quantum states, or pointer states, are selected (Zurek 1998). The selected quantum states are those capable of the greatest reproductive success (Zurek 2004). Clearly Quantum Darwinism is formulated in a manner consistent with the three step algorithm of a Darwinian process.

2) Operation as an adaptive system where the system, the effect of the environment on the system and the effect of the system on the environment are modeled internally and the system conforms to Friston's free energy principle. The simulation of the system's environmental interactions may be understood as computation.

Axiom 1 of quantum theory tells us that all information concerning a quantum system is contained within the state vector making it an excellent candidate for an internal model. As there can be no information without representation this information must physically reside within the quantum system. Axiom 2 tells us that this state vector, and the information it contains, will evolve according to the Schrödinger equation:

$$i\hbar \frac{\partial}{\partial t}\Psi(x, t) = \hat{H}\Psi(x, t)$$

Where $\psi(x, t)$ is the state vector and $\hat{H}$ is the Hamiltonian operator.

As the Hamiltonian operator encapsulates internal effects within the system, the effect of the system on the environment and the effects of the environment on the system this model of the quantum system meets the criteria of an internal model operating within an adaptive system.

Further, we can consider Friston's free energy principle to be satisfied by this adaptive system if we note that Zurek has shown the measurement predictions of quantum theory are implied by the first three axioms. As the predictions of quantum theory are amongst the most accurate in science we might conclude that the prediction error or free energy between the model and the system's actual experiences is minimized.

In Seth Lloyd's treatment of quantum computation (2007) the interactions of quantum systems are interpreted as quantum computations. The result of the computation is the outcome of the interaction. In this sense the state vector precisely predicts the outcome and the prediction error is zero. We note in this context that the noiseless subsystems by which a quantum computation comes to its answer have been shown to be mathematically isomorphic with Zurek's process of einselection (Blume-Kohout, et al. 2008). The isomorphism between these two processes entitles us to view quantum computation as a process where the 'answer' to a computation emerges via a Darwinian process.

Given Lloyd's interpretation that outcomes of quantum interactions are equivalent to simulations carried out by the quantum system, and the fact that all physical interactions are quantum interactions, we are led to the Church-Turing-Deutsch principle which states that every finitely realizable physical system can be perfectly simulated by a quantum computation.

A quantum system, as described by Zurek's theory, meets the criteria of an adaptive system. The internal model formed by a representation of the state vector integrates information concerning the system, the effect of the environment on the system and the effect of the system on the environment. It also operates in accordance with Friston's free energy principle for adaptive systems. The state vector, considered as an internal model, may be interpreted as performing computational simulations of the system's environmental interactions.

3) Maintenance of the internal model's accuracy through Bayesian updating.

In order for an internal model of an adaptive system to reduce its prediction error it must accumulate knowledge through inference from environmental data. It has been demonstrated that there is a single method of conducting inference, Bayes' Theorem, which is derived from the basic sum and product rule of Bayesian probability (Jaynes 1986). Bayes' Theorem provides a method of calculating how confidence in a hypothesis or a model should be updated when new data becomes available:

$$P(H|IX) = P(H|X)\frac{P(I|HX)}{P(I|X)}$$

Where we are considering the Probability of Hypothesis (H) given prior data (X) and new information (I)

In other words we should update our probability for the hypothesis or model being true by the extent to which the hypothesis enhances our ability to predict the new data.

What was formerly axiom 4 of quantum theory reveals that every possible measurable state of the quantum system is predicted by the system state. The correct weighting that should be given to each possible outcome is given by Born's rule. With the occurrence of a measurement of the system the state vector is updated. The probabilistic weightings for the various outcome hypotheses are reassigned in a

Bayesian manner in accordance with axiom 3. The hypothesis associated with a repeat of the previous measurement result is assigned probability one and all other hypothesis are assigned probability zero. In other words, our confidence as to the state of a quantum system immediately after receiving pertinent data is strengthened to certainty.

We can conclude that the internal model of quantum systems undergoes Bayesian updating through the system's interactions with its environment.

4) Construction of adaptations, in accordance with the design specifications of the internal model. The adaptations provide constraints to the system's entropy increase as well as enhanced survivability to the system. These constraints may be understood in terms of the principle of maximum entropy.

Zurek's program has shown that the interaction outcomes involving quantum systems can be predicted by a method named 'predictability sieve' (2003). This method essentially ranks all potential outcomes of the state vector according to their entropy production. Those outcomes that should be predicted (in accordance with former axiom 4) are those having the lowest entropy. This ability of a system to find rare low entropy states which are also reproductively successful is a hallmark of Darwinian processes.

As prevalence towards decoherence is strongly influenced by mass, relatively massive quantum systems such as those considered by atomic physics and chemistry are almost continuously decohered by their environments. Their existence takes the form of what Zurek describes as a classical trajectory, a near continuous decoherence to a classical form. This trajectory is characterized by stability, predictability and reproductive success, and should be understood as a near continuous 'copy with selective retention' process which unremittingly probes potential environmental interactions for low entropy solutions and instantiates those solutions once found. The system's low entropy state, maintained by these interactions is balanced by an increase in environmental entropy (Zwolak, Quan and Zurek 2009).

The evolution of these complex quantum systems, which may be composed of many 'fundamental' particles, is governed by a Hamiltonian operator which models the complex interactions between the system and its environment. The principle of maximum entropy requires that the ability of these low entropy structures to have a well-advertised and persistent presence within their classical environments be explainable in terms of the system's ability to impose constraints on the increase of entropy. These constraints might be considered adaptations and their widespread implementation are interpreted as scientific laws including those governing atomic orbitals and chemical bonds.

Quantum systems evolve by continuously probing their environments to discover and instantiate rare low entropy outcomes at the expense of increasing environmental entropy. The variations adopted by the quantum system in its evolution may be considered adaptations and function to provide constraints against increasing entropy. Much of the subject matter of atomic physics and chemistry can be considered adaptations discovered and instantiated through a Darwinian process.

In this perhaps radical view, quantum systems emerge into classical reality and evolve there through a succession of information transfers between the quantum system and its environment. Each successive generation of offspring information contains variations. The variations which survive and are accumulated over generations are those that succeed in minimizing system entropy. In a suitable environment a quantum system may accumulate atomic and molecular adaptations and function as a relatively complex composite system.

Adaptive systems are characterized by internal models which simulate and orchestrate their environmental interactions. Quantum theory tells us (in the form of a theorem regarding 'no information without representation' (Zurek 2003)), that quantum systems must include a physical implementation of information-processing models equivalent to the evolution of a state vector in Hilbert space. While the physical nature of this model remains outside of experimental verification we are left in a similar situation to biology prior to 1953. At that time many characteristics of life's internal model were known and predictions could be made by invoking laws such as Mendelian inheritance but the manner of the model's implementation in molecular DNA was undiscovered.

In this manner we may view Quantum Darwinism as conforming to the fourth of our listed criteria. This evaluation indicates that Quantum Darwinism meets the listed criteria and may be considered a theory within Universal Darwinism.

It is hard to overstate Zurek's accomplishment in deriving the quantum axioms concerning prediction from those concerning the wave function. Physically his findings imply that the measurable outcomes are modelled by the wave function of the entangled system composed of quantum entity and environment prior to decoherence. The predictive model has two components:

1) The property of the quantum system that will be revealed during the information transfer to the environment will correspond to a 'pointer basis'. These properties are roughly classical in nature and do not include weird superposed quantum properties. The nature of the environmental entities with which the quantum system is entangled will influence the composite wave function so that it models a specific property that will be the subject of the information transferred during decoherence. That property might be position or it might be momentum or something else dependent on the environment.

2) The model predicts the possible values of the information concerning the specified property that will be transferred along with the specific probabilities for each value.

Combined, these two components provide an exquisitely accurate predictive model. Zurek's work demonstrates that this model is not a creation of science, but rather that it is a model inherent in nature which science has discovered (2007). It is this model, contained within quantum systems, which is capable of making the accurate predictions of quantum physics. Human calculations regarding quantum systems amount to simulations of those systems and once we are able to perform these simulations with quantum computers their predictions will be precise.

The predicted probabilities, in a very Bayesian sense, describes the information concerning the quantum entity that will be received or 'known' in the environment of our objective reality. Once an information transfer takes place through the process of decoherence the predictive model inherent in the wave function is updated as required by axiom 3).

Since by definition our objective reality consists precisely of this web of interactions, this is the only information or knowledge available to any of its participating entities. At bottom all information is this type of quantum information.

Quantum theory considered in this context provides a model for what we or any other entity in the quantum web can expect to 'know' about any other fundamental entity in the web. This knowledge is described as a probability distribution over a number of outcomes. It is heavily influenced by prior information; the quantum state at some previous known 'initial condition' and pertinent features of the environment (contained in the Hamiltonian).

A predictive model is inherent in the wave function of the entangled-quantum-entity-plus-environment system which provides probabilities for outcomes. Quantum theory claims to supply the most accurate predictions of expected outcomes that are in principle possible. To date this claim is supported by the evidence. In other words when a measurement, an interaction or a computation is made the 'surprise' or discrepancy between the outcome and the theory's prediction is the minimum possible.

6: Biology

There is a near consensus amongst biologists that life evolved from chemistry on our planet through natural processes. Fundamental to any definition of life are reproduction and controlled local entropy reduction. These characteristics might be taken as a rough demarcation between chemistry and life. There has been much research aimed at identifying plausible chemical systems as candidates for the precursor of life. Gilbert (1986) notes there is extensive evidence in support of various theories explaining how RNA chemistry may have evolved the ability to catalyze its own reproduction.

It is perhaps the central organizing principle of biology that once life did emerge from chemistry it evolved to its present state through the operation of Natural Selection. As Theodosius Dobzhansky, a leading biological researcher of the twentieth century stated "*Nothing in Biology makes sense except in the light of evolution*" (1973).

The emergence of life and its subsequent evolution created the scientific subject matter of biology. The laws of biology, from genetics to population dynamics, may be said to be emergent in the sense that they are not directly derivable from the laws of physics. Natural Selection might be understood as a Darwinian process that, once instantiated in the realm of chemistry/biology, was able to discover survivable biological designs. The design specifications of this emergent biological world form the scientific laws of biology.

Since Darwin published his theory almost one hundred and fifty years ago the evidence in support of it has poured in from numerous disciplines, most notably perhaps genetics. DNA was identified during the 1950s as the molecular unit of storage for much of life's heritable designs. Strong arguments have been made for the biological unit of selection operating at the gene, organism and population levels (Holldobler and Wilson 2008) (Dawkins 1976). It is likely that biological entities are selected at each of these levels. We will examine this process at the organism and population level.

Natural Selection: Organisms

Given that natural selection is the undisputed central mechanism operating in the evolution of organisms we must conclude that this process meets our criteria for the three step Darwinian process.

The design specifications of most organisms are coded in their DNA. A subset of an organism's DNA codes for the assembly of specific proteins which, in concert, compose and orchestrate much of life's chemistry. The successful functioning of this DNA is entirely dependent on the environment in which it finds itself. The organism 'expects' to find itself in an environment similar to that experienced by its ancestors. Crucial components of the genome's expectations include:

1) A specific internal chemistry of the cell including cytoplasm, enzymes and other proteins.

2) Specific sub-cellular bodies and organelles including ribosomes and mitochondria.

3) Structural cellular integrity including cell walls and centrioles.

4) Specific conditions in the external environment including presence or absence of oxygen, pH range, and

available sources of energy which the organism is adapted to consume.

At the time of its conception an individual organism contains inherited DNA that has been under continuous design since the beginnings of biological time. This DNA contains knowledge of the environments experienced by the organism's ancestors. It may be said to represent a model of the environment in which the organism expects to find itself as well as a model of its own structure and functioning. The organism's subsequent interactions with its environment test this model. Large surprises or unanticipated features of an organism's environment for which it is ill suited may result in the organism's death, leaving the field open for those models with variations more closely in tune with the actual environment. In this manner the genetics of individual organisms come to roughly track and model their environments.

All organisms, even the simplest ones, are unimaginably complex chemical entities. In order to exist and to survive their design must include mechanisms for placing effective constraints on the increase of entropy. Unconstrained chemical reactions will spontaneously progress in the direction of increased entropy and decreased free energy. Much of life's design involves placing constraints on this chemical spontaneity. A favoured method of accomplishing this is through the production of enzymes or chemical catalysts which vastly increase the likelihood of specific reactions at the expense of those that would otherwise be more common. Often reactions within the cell are catalyzed to go against the gradient of entropy. To accomplish this outside energy must be made available to the reaction in a highly controlled manner.

Life procures a vast supply of chemical energy for this purpose through the operation of photosynthesis which converts the energy in sunlight into chemical energy and stores it in the chemical bonds of the ATP molecule. In a typical reaction catalyzed to overcome an entropy gradient an enzyme might bind two reactant molecules as well as a molecule of ATP. The reaction is then triggered resulting in an energy transfer from the ATP molecule to a chemical bond between the two reactant molecules. Thus we might conclude that organisms contain many adaptations designed to maintain the organism's survivability and low entropy state in conformance with the principle of maximum entropy and our fourth criterion.

Each gene within an organism's genetic plan codes for a protein which may function as an enzyme directing the course of cellular chemistry or it may take other functional forms. Thus the organism's genetics form its design and serve as a model describing its functioning. Indeed Nobel laureate Sydney Brenner (1999) considers developing scientific understanding of functioning organisms in terms of their genetic models as a central challenge for 21st century biology.

In these terms the genome is a model where genes form a series of hypothesis or expectations concerning the characteristics of the environment in which they will find themselves and the functions they will be able to carry out. Information theory might thus consider an organism's genetics to form a hugely detailed model of their functioning and environment and that individual genes or small groups of genes form sub-models or hypotheses. Organisms' genetics form a model of the expected effects of the environment on the organism and the effect that the organism may have on the environment thus meeting our second criteria.

For example animals within the phylum of vertebrates contain genes coding for the production of haemoglobin, a protein of red blood cells, which collects oxygen in the lungs, delivers it to all cells in the body and receives from them their waste product carbon dioxide which it delivers back to the lungs for exhalation. This gene can be considered a model containing many component hypotheses such as: the animal requires oxygen, the environment contains oxygen, oxygen can be absorbed in the lungs through a certain process, oxygen can be delivered to other cells through a certain process, carbon-dioxide can be captured from cells through a certain process, etc.

Variant genetic codes for haemoglobin may express differences in these hypotheses which in most cases will have less accuracy. The variant haemoglobin may be unable to bind oxygen thus forming the false hypothesis that the organism does not require oxygen. In this case the evidence will confirm the falsity of the hypotheses.

Haemoglobin in humans may come in a variant form known as sickle cell haemoglobin. This variant displays a sickle shape under the microscope which may lead to suboptimal passage through capillaries often leading to reduced health and early death. As humans are sexual their offspring receives one side of their DNA helix from their mother and the other from their father. The debilitating disease sickle cell anaemia occurs when the gene for sickle cell haemoglobin is inherited from both the mother and the father. When a normal copy of the gene is inherited from only one parent the offspring are able to produce enough normal haemoglobin to avoid the most severe forms of the disease.

Interestingly, having one copy of the sickle cell gene and one for the normal gene (heterozygous) provides substantial protection from malaria, such a genetic configuration might be construed as the hypothesis that the person will come in contact with the malaria protist.

These inherited hypotheses are tested during the lifetime of the offspring and the results reflect the nature of the particular environment encountered.   In the above example of heterozygous individuals the hypothesis H, in terms of Bayes' theorem, might be 'I will encounter the malaria protist'. The I or new information might be the degree of selective advantage conferred upon my genetic configuration by the environment. In areas where malaria is prevalent, such as West Africa, up to 25% of the population have the heterozygous form but in other, relatively malaria-free, areas of the world only a very small percentage of the population has this form. X or prior knowledge might be the entire knowledge contained in haemoglobin built up over evolutionary time by my ancestors.

$$P(H|IX) = P(H|X)\frac{P(I|HX)}{P(I|X)}$$

In terms of Bayes' theorem, the crucial component governing the change in the probability of the hypothesis being true is whether the ratio on the right is greater than or less than 1. This in turn is dependent on the whether the truth of H provides an increase in our confidence of experiencing I, the selective advantage which we actually experience.

If H turns out to be correct and the individual does encounter the malaria protist then it is a fact that heterozygous individuals do experience a selective advantage. Thus the numerator of the ratio is enhanced over the denominator which is a kind of average selective advantage over individuals having all types of haemoglobin and encountering the malaria protist.

In this example, which may provide a general model of Bayesian knowledge generation in organisms, the reproductive success of an individual is enhanced for those with the sickle cell gene and who encounter the malaria protist but decreased for those who have the sickle cell gene but do not encounter the disease. This is reflected in the fact that the sickle cell gene is much more common in areas of the world where the disease is prevalent; in both cases the pertinent genetic hypothesis has greater predictive ability. The principle of Bayesian updating of an organism's genetics as illustrated through this example argues for conformance with our third criterion.

Thus we might consider that the theory of natural selection as it applies to organisms meets the criteria we have developed for theories within Universal Darwinism.

Natural Selection: Populations

Population genetics is the study of the change and frequency distribution of alleles under the influence of evolutionary forces (Wikipedia, Population Genetics n.d.). Alleles are gene sequences that may code for a specific protein and that vary amongst organisms within populations of the same species. For instance a population may contain variable alleles coding for a slightly varying proteins which affect the colorations of the organisms. Typically each member of a population will contain a specific complement of that species' genetics, but the population as a whole will contain a frequency distribution of the specific alleles at each gene location on the chromosome.

In a manner analogous to the genetics of organisms the frequency distribution of alleles may be said to represent a model of the environment in which the population 'expects' to find itself. This model has been designed from prior data due to selection pressures operating on the population's ancestors. This prior data, or experiences of ancestors, has been subjected to an inference process (evolution by natural selection), the result being a model which provides expectations concerning the current environment. As the population interacts with its environment this model is tested. Some changed features of the environment may be expected by some alleles but not by others, in which case the alleles that more accurately model the environment may be more heavily represented in subsequent generations. Those alleles that are 'surprised' by a feature of the environment may become less represented or may disappear from the population. This generational copying from phenotype to genotype and back has been described as the central challenge of population genetics (Wikipedia, Population Genetics n.d.).

> *The model contained in the genome of a population will be updated and shift as the population experiences its environment. Those alleles which more accurately model important features of the environment will be more prevalently expressed in the next generation of the population's phenotype. Understanding this generational mapping, in both directions, between a population's phenotype and*

genotype has been seen as the major challenge for population genetics.

According to Lewontin (1974), the theoretical task for population genetics is a process in two spaces: a "genotypic space" and a "phenotypic space".

The challenge of a complete theory of population genetics is to provide a set of laws that predictably map a population of genotypes (G1) to a phenotype space (P1), where selection takes place, and another set of laws that map the resulting population (P2) back to genotype space (G2) where Mendelian genetics can predict the next generation of genotypes, thus completing the cycle. Even leaving aside for the moment the non-Mendelian aspects of molecular genetics, this is clearly a gargantuan task. Visualizing this transformation schematically:

$$ G_1 \xrightarrow{T_1} P_1 \xrightarrow{T_2} P_2 \xrightarrow{T_3} G_2 \xrightarrow{T_4} G_1' \rightarrow \cdots $$

(adapted from Lewontin 1974, p. 12). XD

T1 represents the genetic and epigenetic laws, the aspects of functional biology, or development, that transform a genotype into phenotype. We will refer to this as the "genotype-phenotype map". T2 is the transformation due

to natural selection, T3 are epigenetic relations that predict genotypes based on the selected phenotypes and finally T4 the rules of Mendelian genetics.

Given this framework we can examine the manner in which population genetics operates as an adaptive system according to the four criteria we have developed. First natural selection is central to population genetics and clearly meets the criteria of the three step copy with selective retention mechanism typical of a Darwinian process.

Secondly, the population's genetics form a model of the expected effects of the environment on the population and the effect the population may have on the environment. Population genetics may model a range of expected environments with competing alleles best suited to slightly differing environments. The shifting frequencies of competing alleles over numerous generations reflects shifts in environmental conditions.

Thirdly, natural selection serves to perform a Bayesian update on allele frequencies between generations. A fundamental theorem of population genetics tells us that the frequency of a particular allele in one generation will be its frequency in the previous generation multiplied by the ratio of the particular allele's fitness to the average fitness of all competing alleles. In a population of two competing alleles, A and B population genetics uses an update equation to calculate the expected shift in probabilities. (Ricklefs 1979):

$$p' = \frac{NpR_A}{NpR_A + NqR_B} = p\frac{R_A}{\bar{R}}$$

Where p' is the probability of the particular allele in the latter generation, p is the probability of the particular allele in the former generation, q is the probability of the competing allele in the former generation, N is the number of individuals in the total population, R_A is the fitness of the particular allele and $\bar{R}$ is the average fitness of all competing alleles.

If we compare this with the method used to update a probability with Bayes' theorem we see that the ratio used in Bayes theorem is:

$\frac{P(I|HX)}{P(I|X)}$ and, remembering that this ratio represents the change to our previous state of knowledge realized through gaining new information we see a close parallel between Bayes' theorem and the population genetics update equation. The difference between the numerator and the denominator in Bayes' theorem is that the numerator is the extent to which only a particular hypothesis is predicted by the new information. If the hypothesis predicts the new information then the new information supports the hypothesis and the probability of the hypothesis being true is increased. If the hypothesis did not predict the new information or if it predicted information other than what was received the new information does not support the hypothesis and so our confidence in it is reduced. On the other hand the denominator is a weighted average of the new information being predicted by each of the hypotheses in the family of competing hypotheses.

If we view a competing family of alleles as a competing family of hypotheses that are each modeling an expected encounter with a slightly different situation, then the similarity between these two ratios becomes exact. If the type of expected encounter modeled by allele A is in fact encountered by the phenotype then R_A (the fitness of allele A) is enhanced in relation to $\bar{R}$, the average fitness of the population containing the entire family of competing alleles. Thus the frequency of allele A is increased in the subsequent generation.

In this manner we can view the knowledge accumulated within a population's genome as occurring through the process of natural selection, and that natural selection serves to improve the accuracy of genomic knowledge by performing Bayesian updating on this model as it is transferred between generations.

Lastly, much of the science of biology details adaptations which utilize the principle of maximum entropy to maintain the organism's low entropy state. The family of alleles contained within a population and coded at a particular chromosomal site translate to slightly different proteins, which act as adaptations and serve to tightly constrain cellular chemistry to very specific pathways. These bio-chemical constraints may also be viewed as constraints on the growth of entropy within the organism. A mutant code for a non-constraining protein may lead to reduced fitness or even death, which is the ultimate high entropy state of an organism when its internal constraints cease to function.

Perhaps unsurprisingly we see that population genetics nicely meets the criteria we have set for theories within Universal Darwinism.

7: Brain and Behaviour

The human brain is notoriously complex as indeed are the brains of many animals. It has been estimated that evolution first propelled the information content of the most advanced brains past that of the most advanced genomes about 200 million years ago (Adams 1998).

Unlike the genome there is as yet no overall theory of the brain successful enough to garner near consensus support throughout the research community. While a huge quantity of data has been assembled on brain functions such as sensation, perception, memory, learning etc., there has not been a wholly satisfactory theory explaining them from first principles. However New Scientist reports (Huang 2008, May):

> *That hasn't stopped researchers in the growing field of computational neuroscience from trying. In recent years, they have sought to develop unifying ideas about how the brain processes information so that they can apply them to the design of intelligent machines.*

> *Until now none of their ideas has been general or testable enough to arouse much excitement in straight neuroscience. But a group from University College London (UCL) may have broken the deadlock. Neuroscientist Karl Friston and his colleagues have proposed a mathematical law that some are claiming is the nearest thing yet to a grand unified theory of the brain. From this single law, Friston's group claims to be able to explain almost everything about our grey matter*

This candidate theory of the brain is called the Bayesian Brain theory.

Bayesian Brain

In a recent paper Karl Friston (2007), of University College London, put forward his unified theory of brain functions. This theory is a further development of the Bayesian Brian School of neurology and behaviour which posits that many brain operations follow the logic of inference and may be treated mathematically using the methods of Bayesian probability.

We might tend to assume that humans are the most highly rational organisms but this may be misleading. We should understand that rationality is not necessarily conscious. Any of us that have played games with our pet dogs may well be impressed with their superior ability in many sensory/perceptual/reaction type games such as 'keep the ball from the master'. We may not think of this ability as rational but as Friston shows in this paper all adaptive systems, including a dog at play, must gather information about the external world, infer models of events and causes in the outside world based on this information and take action based on these models. The degree of rationality involved during the 'infer' step will determine the degree of success in interacting with the external world. By this standard our dog is highly rational.

This should come as no surprise. There is after all an external world that exists with very little notice of individual concerns. All organisms are dependent on performing in accord with this external reality for their survival. It is little wonder that mechanisms efficiently (rationally) adapting them to their environments have been built in by evolution from the ground up.

The theory of the brain, which Friston's paper builds on, attempts to explain the brain's underlying rationality as essential to an adaptive system.

Friston begins the body of his paper by defining the general characteristics of an adaptive system as one that can react to changes in its environment by changing its interactions with the environment to optimize the results for itself.

Most of us would view his definition as a fair one. Certainly any system that operated according to the definition should be considered 'adaptive'.

He examines an adaptive system in term of three features or variables: the system itself, the effect of the environment on the system and the effect of the system on the environment. Friston then expands on his definition and claims that it is equivalent to the one stating that the system attempts to change the effect of the environment on the system either by changing itself or by changing its effects on the environment. He then makes a further shift and claims that a third definition is also equivalent: an adaptive system is one that minimizes unlikely or surprising exchanges with the environment. In other words the best strategy for an adaptive system is always the one in which the environment's effect on the system was expected. We might consider this to mean where the system's internal models are 'in tune' with external reality.

It does not seem to me that the logic of the third definition is equivalent to the logic of the first. The third definition talks only of making an internal model (expectations) as close as possible to what actually happens. What happened to 'optimizing for our benefit' and what happened to the effect we have on the environment? I assume Friston takes it for granted that those issues have been dealt with by other processes and that both the expectation we have and the actual outcome have already been optimized, the actual outcome by the internal changes we have made to our system, or by changes we have made on our effect on the environment. Yes, in this scheme we can and do manipulate how the environment affects us so we can optimize what actually happens. Given this assumption I think his third definition holds.

He is able to express the logic inherent in this third definition as a mathematical function that takes the form of Bayesian probability, if he introduces one more variable to denote unknown environmental forces that cause the effect the environment has on the system. Once he has done that, he moves quickly to deduce a physical characteristic of adaptive systems that must be maximized in order for his third definition to hold: free energy. Unfortunately that formula contains the new variable, the unknown environmental causes. On first consideration this may look deadly as how could the adaptive system operate in this manner if it has to have as a precondition the answer to the very puzzle it is trying to solve (adapt to)?

Amazingly, with a little mathematical sleight of hand Friston escapes this conundrum. He shows that while the system is unable to calculate the exact free energy function it is able to compute a limit or bound on that function. That bound can be computed without any reference to the new variable (unknown causes in the external world). After some further mathematical contortions a link in the form of a probability density is derived that connects states of the adaptive system to the unknown environmental causes. In other words he shows that a necessary characteristic of an adaptive system is the possession of a model of the outside world.

In Friston's words:

> *The free-energy formulation in Eq. 3 has a fundamental implication: systems that minimise the surprise of their interactions with the environment by adaptive sampling can only do so by optimising a bound, which is a function of the system's states. Formulating that bound in terms of Jensen's inequality requires that function to be a probability density, which links the system's states to the hidden causes of its sensory input. In other words, the system is compelled to represent the causes of its sensorium. This means adaptive systems, at some level, represent the state and causal architecture of the environment in which they are immersed. Conversely, this means that causal regularities in the environment are transcribed into the system's configuration.*

Perhaps the coolest thing about this theory is that the internal model that must be developed on the basis of mathematical inference from sensory data is exactly analogous to science. The theory posits that the brain unconsciously smoothes the system's challenges with the environment by practicing the scientific method and building accurate models or theories concerning the operation of its environment through making rational inferences from sensory data. When we turn our head and/or eyeballs to better focus on that thing we saw in our peripheral vision our brain has unconsciously taken action to gather data with which to decide how best to weigh the probabilities of the various competing hypotheses contained in its model.

Thus Friston arrives at a powerful, mathematically tractable model of brain function. It utilizes inferential logic (Bayesian probability) and with his Free Energy principal he is able to compute and predict a wide range of measureable neurological results.

This model of brain function can explain a wide range of anatomical and physiological aspects of brain systems; for example, the hierarchical deployment of cortical areas, recurrent architectures using forward and backward connections and functional asymmetries in these connections (Angelucci et al., 2002a; Friston, 2003). In terms of synaptic physiology, it predicts associative plasticity and, for dynamic models, spike-timing-dependent plasticity. In terms of lectrophysiology it accounts for classical and extra-classical receptive field effects and long-latency or endogenous components of evoked cortical responses (Rao and Ballard, 1998; Friston, 2005). It predicts the attenuation of responses encoding prediction error with perceptual learning and explains many phenomena like repetition suppression, mismatch negativity and the P300 in electroencephalography. In psychophysical terms, it accounts for the behavioural correlates of these physiological phenomena, e.g., priming, and global precedence (see Friston, 2005 for an overview).

It is fairly easy to show that both perceptual inference and learning rest on a minimisation

of free energy (Friston, 2003) or suppression
of prediction error (Rao and Ballard, 1998).

This theory does seem to explain a great deal of the data that has been gathered on brain function and has energized and excited many of the researchers in the field.

In considering whether the Bayesian Brain theory is a candidate to be included within Universal Darwinism we must rate it against our four criteria. Indeed the last three criteria are pretty much straightforward:

2) The Bayesian Brian theory obviously complies with the criteria that it treats brain operations as an adaptive system where the system, the effect of the environment on the system, and the effect of the system on the environment are modeled internally, and the system conforms to Friston's free energy principle.

3) It is a fundamental tenet of the theory that the internal model's accuracy is maintained through Bayesian updating.

4) Remembering that the Free Energy principle of the Bayesian Brain theory requires that the brain operates so as to reduce expected surprise or entropy, we may view an organism's neural-based adaptive behaviours as adaptations which utilize the principle of maximum entropy to maintain the system's low entropy state.

However a search for a three-step Darwinian algorithm of 'copy with selective retention' within the literature of the Bayesian Brain theory comes up empty. The only mention of a Darwinian process seems to be natural selection and refers to genetics.

This might be expected as the Bayesian Brain theory is not an attempt to explain brain function in biological terms. It is a theory which models brain function computationally and mathematically but does not dwell on the physical implementation actually occurring within the meat of the brain.

Fortunately there are a number of theories, consistent with the Bayesian Brain, which provide models of Darwinian processes and which may be viewed as describing physical mechanisms for carrying out the computations described by the Bayesian Brain. These theories include Neural Darwinism proposed by Nobel Laureate Gerald Edelman (1987), the Natural Selection in the Brain theory of Eörs Szathmáry (2008) and the Synaptic Darwinism theory of Paul Adams (1998).

Synaptic Darwinism

Donald Hebb, the Canadian psychologist sometimes called the father of neural psychology and neural networks, published the theory of Hebbian learning in his book Organization of Behaviour (1949). This theory provides insight into many psychological processes such as learning, memory and thought and has endured as a fundamental component in our understanding of the brain.

Starting with Pavlov in 1927 it had been noted that learning often took the form of conditioned response. If a stimulus were repeatedly paired with a response then the behaviour would become learned and the behaviour would tend to be elicited by the stimulus alone. For instance Pavlov noted that dogs salivated when given food and correctly predicted that if a bell was repeatedly rung when a dog was presented food then following a number of repetitions the dog would salivate whenever the bell was rung even when no food was presented.

Research into this type of learning dominated neural psychology during the middle decades of the past century. Hebb's major contribution was to propose anatomical mechanisms which might produce learning and other associated mental constructs such as memory and thought.

Central to his program is the notion of Hebbian learning which states that if two neurons are repeatedly activated at the same times then the links between them will be strengthened. In Hebb's worlds (1949):

> When an axon of cell A is near enough to excite B and repeatedly or persistently takes part in firing it, some growth process or metabolic change takes place in one or both cells such that A's efficiency, as one of the cells firing B, is increased

This mechanism of Hebbian learning provides an anatomical explanation for conditioned response. If for instance cell B is involved with a response such as salivation and cell A is involved with a stimulus such as the ringing of a bell then the learning of salivation in response to the ringing of the bell is explained as an increase in the ability of cells in the group which includes cell A to cause firing in the group of cells which include cell B.

In the decades since Hebb many further anatomical details of learning have been discovered which fit nicely within this theory. Recently Paul Adams has proposed the theory of Synaptic Darwinism to provide a more complete explanation of the workings of Hebbian learning (1998). In this theory the 'growth process or metabolic change' which takes place between the two neurons is a copying of synapses joining the neurons.

This copying is not precise and sometimes a synapse is copied joining Cell A with some neuron with which it had not been previously connected, providing a measure of variability in neural connection patterns.

Adams details neurological machinery that is able to compare rewards, for instance food presentation, with other environmental cues, such as bells ringing. The comparison of cues with rewards provide the basis of a selection process where those neurons associated with cues related to rewards are provided resources for replication while those neurons associated with cues not involved with rewards are denied resources.

Thus Adams's theory nicely takes the form of a Darwinian process with a three-step algorithm: copy, variation, selection.

In a personal communication Paul Adams indicated that Synaptic Darwinism is completely consistent with the Bayesian brain theory. While the Bayesian brain theory has a focus, within the constraints presented by the data, on the mathematical and computational structures of brain functions, the theory of Synaptic Darwinism offers an explanation of the anatomical details through which they may be implemented.

This is an active area of research and the theory of Synaptic Darwinism has numerous competitors including the theories of Natural Selection in the Brain and Neural Darwinism. However it seems extremely plausible that whichever theory eventually gains ascendance its explanation of anatomical brain functions will be an explanation utilizing a Darwinian process. By putting such a Darwinian process within the context of the Bayesian brain theory we arrive at the conclusion that the composite theory meets our four criteria as a candidate theory within Universal Darwinism.

8: Culture

The evolution of human cultures over the past 100,000 years represents the most dramatic explosion of low entropy design known in the history of the universe.

Culture is clearly an entity emergent from the biological realm whose evolution is subject, at least in part, to its own emergent laws.

The prior knowledge from which culture emerged is huge, and encompasses both chemistry and biology. Numerous subconscious mental processes, illuminated by the Bayesian Brain theory, provide components used in cultural processes; of these perhaps consciousness is the most essential.

Consciousness, specifically that denoted by *qualia* (the subjective experience of consciousness, for example the experience of "redness" by a visual system) is deemed 'the hard problem' by philosophers (Shear 1999). From a functional point of view, consciousness may have arisen due to a need to bring a kind of "meta-order" to mental mechanisms that are rapidly evolving in complexity. At least the degree of consciousness in animals seems to parallel their mental complexity as exemplified by humans. Some recent research suggests that the purpose of consciousness and qualia may be to flag (and thereby differentiate between) our experience of outside reality and internal modeling. This gives a special quality to the experiencing of the current sensory information we are receiving from the environment so that we don't confuse it with mental models such as memories, plans and day-dreams (Gregory 1998).

Try looking intensely at some distinctively coloured object, such as a red tie. Then close the eyes and imagine the tie. The vivid qualia are suddenly far dimmer in imagination. To reverse the experiment, imagine the object, then open the eyes and look at it. The qualia of the visual now are startlingly vivid by comparison with the memory. So perhaps what qualia do is flag the present so that we do not get confused with a remembered past or anticipated future.

Another widely accepted precursor of culture is the related abilities to imitate and learn. Cultural processes evolve over time as they are passed from generation to generation through a process of learning or imitation. Humans excel at the ability to imitate and some detailed theories of this ability are under development (Iacoboni 2005):

> *Brain imaging techniques allow the mapping of cognitive functions onto neural systems, but also some understanding of mechanisms underlying behaviour. A series of imaging studies have described a minimal neural architecture for imitation. This architecture comprises a brain region that codes an early visual description of the action to be imitated, a second region that codes the detailed motor specification of the action to be copied, and a third region that codes the goal of the imitated action. Neural signals predicting the sensory consequences of the planned imitative action are sent back to the brain region coding the early visual description of the imitated action, for monitoring purposes ("my planned action is like the one I have just seen"). The three brain regions forming this minimal neural architecture belong to a part of the cerebral cortex called perisylvian, a critical cortical region for language. This suggests*

that the neural mechanisms implementing imitation are also used for other forms of human communication, such as language. Indeed, imaging data on warping of chimpanzee brains onto human brains indicate that the largest expansion between the two species is perisylvian.

The centrality of imitation to culture is conveyed in its Wikipedia article: (Wikipedia n.d.):

Culture can be defined as all the ways of life including arts, beliefs and institutions of a population that are passed down from generation to generation.

Given that imitation is central to culture it is easy to see that cultural processes might often involve Darwinian processes:

1) Copy – imitate

2) Variations in the copies - learning and imitation do not involve perfect duplication but produces some variable perspectives.

3) Variations in the characteristics of copies influence their survival - sometimes the new perspective is superior or better adapted and becomes more widespread; sometimes the new perspective has inferior survivability and is not long retained.

This model has been widely adopted in the social sciences, and fields of study with 'Evolutionary' in their title abound: Evolutionary Psychology, Evolutionary Archaeology, Evolutionary Linguistics, Evolutionary Epistemology, Evolutionary Economics etc. The school of Memetics views cultural evolution in general as based on a Darwinian process involving memes as the unit of replication (Dawkins 1976). Another general theory of Darwinian cultural evolution is the dual inheritance theory (Henrich 2007) .

The fields of study listed above all employ Darwinian processes, with the three step 'copy with selective retention' algorithm as the engine of their evolution. The diagram below is from a paper in evolutionary archaeology and illustrates the adoption of the biological method of cladistics, with only minor alterations, to the study of arrowhead design evolution amongst native peoples of south-western North America (O'Brien 2003). Cladistics is used extensively in biology to illustrate the evolutionary relationship of life forms. The adoption of this method by Evolutionary Anthropology supports the assumption that the same type of Darwinian evolutionary mechanisms underlies the evolution of design in cultural artefacts such as arrowheads.

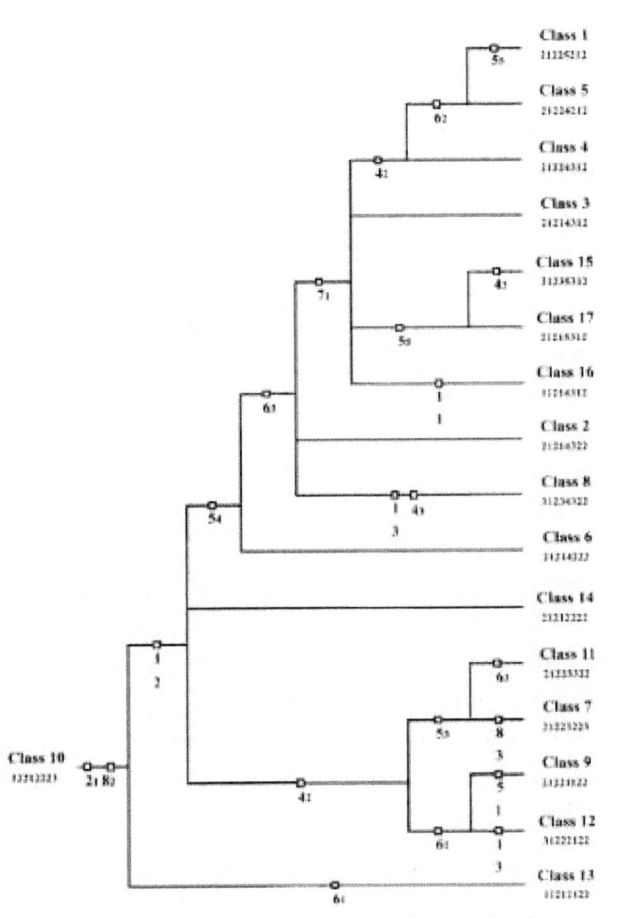

Figure 9.6. Phylogenetic tree of 17 projectile-point classes. The tree has been slightly simplified from the original (see O'Brien et al. [2001] for details). Class 10, shown here as an ancestor to all other classes, is the outgroup against which all other classes were compared. In a strict cladogram, it would be shown as a terminal taxon (located at branch tips), as are all the other taxa.

Science

Two hundred years ago about 1 billion humans inhabited the earth. Even in the most advanced countries the average person could hope for a life expectancy of forty years and wealth of about four hundred dollars per year. Today we are almost seven billion, have an average life expectancy that is approaching seventy years and average income approaching thirty thousand dollars a year. An amazing YouTube video, called Joy of Stats, featuring researcher Hans Rosling demonstrates this data. After struggling for success as a species for 100,000 years, in the blink of an eye, in two hundred years, we have attained the Holy Grail coveted by all species since the beginning of life on earth with our reproductive success, our survival rate and the resources we command. At the heart of this success, setting us apart from all other species, is science.

Having gained a foothold only in the mid 1600s, science required a lengthy incubation before giving birth to this transformation. Once its inexorable process of knowledge accumulation got underway science informed many other cultural adaptations such as medicine, agriculture, transportation, public health, mining and the production of goods. The injection of scientific knowledge into these adaptations acted as a steroid powering them to a trajectory of exponential growth, complexity and productivity.

Today six billion people, those in excess of the 1 billion who inhabited the earth two hundred years ago, owe their existence as well as their unprecedented health and wealth directly to the power of science. Take science away from humanity and most of the earth's human population would cease to exist. Those remaining would experience a much shorter life and would command a relatively paltry amount of resources.

The extent to which we take this transformative source of our species' good fortune for granted is perhaps even more remarkable. Any widespread appreciation of the astounding creativity of science or even an acknowledgement of its vast and obvious power seems to have eluded most of us.

While we demand ever more from its resource-producing and life-extending abilities we balk at heeding any scientific advice concerning the misuse of scientific power. We routinely discount the counsel offered by science and seem instead to prefer myths featuring stern father figures who take a personal interest in us.

However culture is in the same predicament as all other evolving entities; complexity requires greater knowledge and greater knowledge results in more complexity. Although a species which spent most of its evolutionary history on the savannahs of Africa may not find this predicament comfortable we will eventually have to face the reality that being at the cutting edge of evolution and clinging to the comforts of ignorance are not compatible options.

Scientific knowledge is largely the same as the other forms of knowledge found in nature. It evolves through a process of inference from information and forms models which accurately predict the outcomes of events in the environment. Science is largely concerned with the nature of other complex entities which in turn are based on knowledge, knowledge that evolved to its current state through the process of inference. Perhaps the one unique aspect of science is that its models aspire to a complete representation of the universe and the entities within it.

Science, as seen through the lens of Universal Darwinism, is a human rediscovery of the knowledge developed by nature over evolutionary time since the big bang. It provides explanations for the functioning of the various emergent entities which have assembled the knowledge necessary for existence.

The power of science is due to the immense prior knowledge inherent within the human organism. We encompass, and function through accessing, the knowledge of quantum physics, chemistry, biology, neurology and culture. This vast storehouse of knowledge, built up over evolutionary time through the process of inference, allows us in turn to perform more direct and powerful methods of accumulating knowledge, also via inference. Science is the most direct method yet discovered.

I will argue, as have some eminent philosophers of science that the methodology of science is a Darwinian processes, that science conforms to the same paradigm as does all other known forms of knowledge accumulation. In this view science is but another method (admittedly uniquely direct, comprehensive and powerful) employed in nature's on-going quest of exploration and discovery.

Explanation, Evidence, Truth and Power

Introduction

We often hear that the truth is unknowable or that what is true is a matter of opinion. As the possibility of knowing truth is often doubted we are tempted to settle for beliefs due to their psychological appeal. If reality is unknowable this would be a rational strategy. If reality is so ridiculously complex that all true aspects of it are beyond our understanding, we should not waste our time pursuing truth. If this is not the case and we do have the capacity to understand significant aspects of our true circumstances then we should value those systems that are able to convince us of their truth.

As discussed in previous chapters we are knowledge machines from the ground up. I believe that our capacity to know significant truths is a closed question. Nearly all cells conduct the chemical symphony that creates energy from glucose. Some reasonable definitions of the word 'to know' support a usage such as: 'A cell knows how to convert glucose to energy.' A definition including this kind of knowing, dissociated from anthropomorphic conscious processes, would also support usages: 'A bird knows how to fly' or 'A monkey knows how to keep its balance.' Daniel Dennett has argued that consciousness may only be a negotiation mechanism to arbitrate amongst the myriad number of unconscious knowledge mechanisms.

The definition of knowledge we are developing here includes unconscious knowledge of the kind noted above. According to this definition the world is replete with instances of significant knowledge. We are in essence entities that know a method of surviving in the circumstance in which we find ourselves. This knowledge has been embellished, honed and passed down to us through each generation of living things since the beginning of life. As Henry Plotkin argues in <u>Darwin Machines</u>, a work that has transformed evolutionary psychology, intellectual forms of knowledge are simply new adaptations evolved from earlier ones (1993).

Man incorporates all types of knowledge produced by Universal Darwinism. Knowledge associated with atoms, complex chemistry, biology and culture are all components of human beings. No other known being incorporates this full range of knowledge.

In a sense the relentless production of knowledge by Universal Darwinism represents an attempt by the universe to come to know itself. Science is arguably the most highly evolved mechanism providing this self knowledge. The details of the working of science provide insight into its ability to discover truth and bestow power.

Science, truth and power

For our purposes we might define power as the ability to cause the occurrence of phenomena at will. This definition implies both physical and mental components of power. The mental components are involved in 'our ability to cause something to occur at will'. The part about 'causation of phenomena' implies some physical components.

People have a predilection to understand power mostly from the mental perspective. We seem to have built-in mechanisms for attributing human-like motivations to many aspects of nature and we yearn to relate to nature in this manner (Boyer 2001). People often attempt to invoke power through purely mental means. We implore ghosts, witches, saints and all manners of spirits to intercede on our behalf. This has gone on for millennia despite having virtually zero measurable effect. Still it seems our first impulse when attempting to invoke power.

Unfortunately matters of survival seldom depend on the purely mental. There has to be a physical component, for survival takes place in the physical world and those of us here today are descended from extremely long lineages of beings that were able to make the right choices concerning survival in this arena.

Perhaps a clear example is war. War is an arbiter of power in that it clears the field of obstacles impeding one's ability to expropriate the resources of an area. With these resources in hand one can cause lots of phenomena to occur. War is definitely about survival in the real world, it can be a grinding, real, physical event that produces a clear outcome. It can leave a group with clear possession of some real estate.

War can also arbitrate ideas and mental traditions; it can be a deciding event in the conflicts of cultures. Many cultures have ceased to exist following a conquest.

War also involves the mental: tactics, strategy and technology. These are mental constructs crucial to the outcome. Often the day goes to him who has these kinds of issues well thought out. Power is never wielded by the mental alone; trying to influence spirits won't do it. Only when the mental is in tune with the physical does it actually produce power.

Technological abilities may not be included in many people's catalogue of abilities required by successful generals. Usually we think of abilities such as tactics and strategy. This is probably only confusion in the use of the word 'technology'. If we had used the word 'weapons' instead no one would quibble unless with the implication that weapons are a mental component. But weapons are highly dependent on the mental. Whether one learns their particular culture's art of flaking arrow heads or developing smart bombs, weapons require understanding of the world; they are a type of technology.

With science weapons conveying colossal power have been devised. The energy of modern weapons produce extremely intense phenomena; they are vastly powerful. What is the source of that power? How, after hundreds of millennia of fruitlessly invoking spirits, have we in the last few centuries stumbled, so ill prepared, onto this vast source of power? How did it happen?

Well, it happened because we have unlocked a new means of putting the mental in tune with the physical; a new (at least to us) means of discovering truth and unlocking power.

A brief example will help to clarify the relation amongst scientific explanation, truth and power. At the turn of the century the foremost physicist of the day, Ernst Rutherford, believed he had disproved the theory of Darwinian evolution. He reasoned that the sun, if it derived its energy from any plausible physical source, would have burned out long before the billions of years required by Darwin's theory to produce complex life on earth. Therefore the sun and earth could not have been in existence for the great length of time required by Darwin's theory.

A scientific explanation of nuclear energy had not yet been developed. Some data had been collected by investigators such as the Curies, but nothing resembling our current explanation was available. In 1905, as part of relativity theory, explaining the weird fact that the speed of light is the same for all observers, Einstein derived the formula $E=mc^2$. The main take-away from this formula is that a little matter is equivalent to massive amounts of energy. By the early thirties, a lot of data concerning the behaviour of stars had been gathered. It was known that they were composed mostly of hydrogen, with older stars having a higher proportion of helium than younger stars. In 1939 Hans Bethe explained that stars were fuelled by nuclear fusion where four hydrogen atoms fuse to form one helium atom. Four hydrogen atoms have slightly more mass than one helium atom. In a star the excess mass is converted to energy exactly according to Einstein's equation. Bethe's theory explained all the scientific data collected concerning stars and nuclear reactions. The sun was seen to have had a multi-billion year history and this challenge to Darwin's theory was vanquished. Within twenty years, using this theory, the hydrogen bomb had been developed and tested. We could create a mini sun here on earth, at will.

Human history has probably developed thousands of explanations, many of them religious, for the sun's existence and functioning. Only the one developed by science conforms to the evidence, only science successfully puts the mental in tune with the physical and unlocks powers, such as that of the sun.

It is the knowledge developed by science that contains the ancient truths. The phenomenon of suns, replicated in scientific knowledge, has been in the universe since there first were suns. We have only recently developed scientific constructs, mental constructs that are in tune with this physical phenomenon. Science is mainly discovery, not invention. The meta-theory of Universal Darwinism is composed of scientific theories spanning the range of evolutionary time and brings a focus to the many knowledge entities which have evolved during the history of the universe.

Science evolves toward truth

The theory of memes claims that all cultural traits are learned from others or are imitations of behaviours and are created and modified by the processes of evolution. All systems of knowledge evolve as they follow the Darwinian algorithm. In this sense science is not special and shares the field with other systems of knowledge such as theology. Each theologian accepts a shared body of knowledge, which is reproduced in her mind with some variations from the consensus beliefs. Some of these variations may prove to confer survival value on the belief and are passed on to other believers. Thus the body of theology evolves.

But what gives direction to theology's evolution? What are the criteria that lend survival value to certain variations? A major criterion must simply be conformity to the believers' (especially the more influential believers') religious intuition or other types of subjective advantage. In the view of meme theory theological adaptations have survival value depending on how well they mesh with and how attractive they are to the individuals' existing meme complexes. Confirmation by objective evidence plays little role in the survival value of theological variations.

The difference between the evolution of science and religion lies in the criteria that defines fitness of adaptations and therefore determines what will survive and be passed on. Science demands that its knowledge conform to the evidence. Religion does not. Faith, an integral part of many religious systems, can be defined as a commitment to believe regardless of the evidence.

Evidence is important because there are reasons to believe it is our best guide to what actually takes place in reality. Scientific evidence is evidence confirmed by the senses. Our senses provide ways of knowing honed to a high degree of accuracy by evolution over geologic time. Knowledge that cannot bear the scrutiny of the senses is rejected by science.

This process gives direction to the evolution of scientific knowledge. Scientific explanations evolve to more closely conform to data confirmed by the senses. Over time they tend to agree to more decimal places. As science unlocks the secrets of the material world observed by the senses it becomes more powerful. In some sense it is inescapable that the tremendous power of science lends credence to its claim to be a 'true' system of knowledge.

Scientific theories are rational stories whose accuracy is verified by experiments. Experiments produce phenomena in the real world that may lend credence to a theory. A valid experiment must be verifiable or able to be reproduced by other researchers. Experimental phenomena that can be reproduced by other researchers may also be produced by technicians at will. This explains the power of science; its ability to cause phenomena at will. Experimental evidence binds science to the 'real world' and is the source of its power. Scientific knowledge providing highly accurate models honed through inference is knowledge that is able to deliver power.

The scientific explanation of the sun's working evolved a great deal in forty years following Rutherford's challenge to Darwin. It grew in tandem with a growing body of scientific evidence, evidence composed of observations revealing data to the senses. At every step in its evolution the explanation was constrained to fit the available evidence.

It seems clear that the scientific explanation of the sun's workings evolved closer to the truth. One group of philosophers, deconstructionists, deny this. They contend that science is only one amongst many cultural bodies of knowledge and has no special claim to the truth. Paraphrasing Dr. Johnson, deconstructionists, along with anyone else, are vaporized in the presence of atomic blasts and are thus refuted. In some sense science is not just another body of knowledge but one that can claim to have unlocked truths and powers of the world unmatched by any other system of knowledge.

A nagging question remains. Why should we prefer an accurate or 'true' system of knowledge over one that is less so. Why should 'true' explanations have survival value? Perhaps this preference is not new but is built upon the survival value of accuracy in sensory perception evolved in our ancestors from the earliest times. Daniel Dennett makes this point (1995):

> *Getting it right, not making mistakes, has been of paramount importance to every living thing on this planet for more than three billion years, and so these organisms have evolved thousands of different ways of finding out about the world they live in, discriminating friends from foes, meals from mates, and ignoring the rest for the most part.*

Our bodies reward our conscious minds when we behave in ways the body approves of and the body punishes us when we don't. We experience pleasure and pain as part of a feedback mechanism designed by evolution to keep us on the straight and narrow. It's why sex isn't a pain.

On the pleasure side, endorphins are released into the blood stream at appropriate times and stimulate the brain's pleasure centres. We gain pleasure from many things, good food, sex, family and creativity. Einstein's claim that the cosmic religious experience is at the root of all science (1930) provides a testament to the pleasurable mental state induced by creative science and goes some way towards explaining the motivation of those who create science as well as those who strive to understand and appreciate it.

Science is built with tools developed by evolution.

Perhaps the most distinctive human biological characteristic is our huge brain. This large size comes at an extravagant cost. It poses increased risk during childbirth and consumes up to 20% of the bodies energy production. Explaining why we have a brain of this size and cost is a challenge for evolutionary theory. What is its survival value? Explanations based on biological evolutions are only partially convincing; it is difficult to imagine our remote ancestors receiving a biological advantage large enough to provide the required enormous differential survival.

I find a recent suggestion by Susan Blackmore (1999), based on meme theory, to be more convincing. She suggests that our large brain developed due to its ability to manipulate memes. Memes, a concept introduced by Richard Dawkins (1976) are:

> *tunes, ideas, catch-phrases, clothes fashions, ways of making pots or of building arches. Just as genes propagate themselves in the gene pool by leaping from body to body via sperms or eggs, so memes propagate themselves in the meme pool by leaping from brain to brain via a process which, in the broad sense, can be called imitation.*

Blackmore postulates that selective pressure for large brains increased due to greater biological survival of those individuals best able to learn and manipulate memes. She takes the concept further and identifies memes as a second replicator, genes being the first. Evolution of many memes, such as science, are not strongly tied to biological survival, but follows their own logic of fitness.

Whatever their exact origins, we have both a large brain developed by evolution and a mature system of knowledge, also developed by evolutionary processes.

Parts of the human brain are very similar to the brains of those species who are our closest relatives, earth's other primates. In particular our sensory systems are very similar. Other parts of our brains are new and largely unique to humans, the parts that handle higher intellectual functions and memes. The explanations of science are clearly stored and manipulated in these new areas of the brain as are all other systems of knowledge. The distinctive characteristic of science, as a system of knowledge, is that it also utilizes functions of the old brain. The construction of new variant scientific explanations is performed in the meme-handling centres of the new brain. Judgements resulting in the differential survival of these variants are formed on the basis of sensory data supplied by the old brain. New variants live or die, are passed on or ignored, depending on evidence revealed to the senses of the old brain.

In terms of meme theory, the most accurate knowledge tools produced by both types of replicators are integrated into scientific methodology. Scientific explanations, or memes, are replicated with variations. The fitness of these memes are judged by evidence supplied by the most accurate knowledge tools produced by genetic replicators; the senses. This synergistic utilization of functions from both the old and new brains powers scientific evolutionary progress. This synergy allows scientific explanations to be demonstrated in the real world for all to see. Seeing is believing. Scientific truth is thus believable in a manner unparalleled by other system of knowledge.

Sense perceptions (functions of the old brain) straddle the conscious/unconscious boundary. On the unconscious side resides knowledge such as that involved with vision, which translates photons and patterns of photons into neural signals meaningful to other parts of the brain. On the conscious side are mechanisms for providing these signals with context and for altering behaviour according to the content of the signals. Sense perceptions have been evolving for eons as a useful way of knowing. They have given power to those who possess them and are almost universal within the animal kingdom. Competitors lacking highly evolved senses are unable to compete in almost all circumstances.

Clues abound that sense perceptions provide us with very reliable knowledge as compared with the speculative nature of higher intellectual mechanisms. For example the Law rightly has a preference for an eyewitness over someone who witnessed the event in their imagination. A fact may be defined as something having objective reality or having been confirmed by the senses.

The interpretation of sensory data may sometimes lead to discrepancies. A group of eye witnesses may disagree as to significant aspects of an event. The claim here is not that the senses are perfect, only that they are our best and most trusted guide to many aspects of reality.

Science has been described as a mechanism for developing explanations that best fit the scientific evidence (Deutsch 1997). Scientific evidence is largely produced by experiments designed to expose some aspect of reality to the inspection of the senses. Measurements are made, readings are taken, and information is received in the conscious mind from the senses. These data points constrain the artistic freedom of the explanations and ultimately pronounce judgement on them.

Pre-scientific Greek philosophy extolled the virtue of pure thought over sense impressions. Aristotle argued that men had more teeth then women. This was accepted as the truth by European experts for nearly a thousand years. Aristotle arrived at this and many other beliefs through a process of pure thought; it made sense to his imagination. Truth existed in the human mind, everyday reality was a poor imitation of the ideal world and not worthy of study. Not until a more scientific era did it seem appropriate to actually look in peoples' mouths, count their teeth and decide that in fact men and women had the same number of teeth. Movements for human equality may well best flourish during scientific eras, when some weight is given to evidence.

Evolution of life forms on earth has been going on for nearly four billion years. An astounding variety of complex designs have evolved during this vast expanse of time. That this was accomplished by natural selection without a plan or a designer is a central tenet of biological evolution. Natural Selection is based on the differential survivability of inherited characteristics. Each generation inherits characteristics from their parents and in each generation there is some random variation in these characteristics. Some variations will bestow greater reproductive success on their bearers than will others. The individuals possessing these variations will propagate more offspring, offspring tending to have these parental characteristics as well as exhibiting some variations of their own. In this manner each generation tends to accumulate adaptations that bestow reproductive success.

There is no plan. What can persist does persist. What cannot persist passes away. In a sense it is very wasteful to have a great variety of designs tried out in each generation only to have most of them discarded. Wouldn't foresight give better odds for survival? It would and evolution produced it when it produced consciousness in animals. Consciousness allows us to simulate actions or courses of behaviour in our imaginations and see how they might play out. Would attacking that Sabre-Toothed Cat really be a good idea? Many courses of action endangering survival do not need to be actually tried out, we can see right away they are bad ideas and try to think of something different.

Of course foresight is good only if it is true to its name and is somewhat accurate. Imagining totally inaccurate outcomes would be useless. It is clear that our imagination is not totally accurate; Aristotle could deduce different dental configurations for men and women. We are quite often surprised by things turning out different then we foresaw. Fortunately our conscious simulations are much better than nothing. Often we foresee important events and are able to take steps to optimize our situation.

Science is interplay between our senses and higher intellectual faculties. The imagination explores variant explanations which must live or die according to their fitness. Their fitness is in turn determined by their ability to explain evidence gathered by the senses. Science, a higher intellectual system of knowledge, explains how things work in the 'real' world, the world as revealed to our senses. In this manner its explanations conform to the evidence, have power in the 'real' world and are thus confirmed to be 'true'.

Bayesian Probability: The Logic of Science

Science is a process integrating rational explanation with empiricism. A scientific theory lives or dies on its ability to rationally explain empirical data. This may seem a bit fuzzy as plausible theories of nearly infinite variation may be constructed concerning some given subject matter as long as none explicitly contradict the data. For instance all explanations that do not relate to the collected data cannot be ruled out by the data. How are we to decide amongst them on the basis of data? Bayesian probability, in the form of Bayes' theorem provides a mathematical framework for measuring the quantitative fit between empirical data and the theories competing to explain the data.

Pierre-Simon Laplace, a great scientist of the 16th and 17th centuries, was an early developer of Bayesian probability and used it most effectively to fill in the details of Newtonian celestial mechanics. Celestial mechanics is the study of the motion of bodies within the solar system and has as its main theoretical underpinnings Newton's second law of motion: F=MA. This famous physical theory relates the mass and acceleration of a celestial body to the forces acting upon it. In the case of celestial mechanics the force is Newton's theory of universal gravitation and the acceleration of the body is usually described by its orbital path. The mass is a constant and is a property of the specific body. Data collected to test Newton's theory usually focus on measuring the position of a celestial body at successive times. The body's acceleration is the deviation of the body's motion from a constant speed in a straight line. The body's path is then predicted as a calculation directly from Newton's theory relating the acceleration of the body to its mass, the mass of all other relevant bodies in the solar system and the body's distance from them. The two results are then compared and close agreement of the data with the theoretical prediction is taken as experimental support for the theory.

Unfortunately the masses and distances required to calculate the theoretical predictions are only known within a margin of uncertainty. Measurements of these quantities give a definite answer but successive measurements do not give exactly the same quantity; there is some uncertainty in the measured value. This uncertainty usually is well described by the bell curve, where the values of most measurements are clustered around some central value while measurements deviating from that central value will be less common the further they deviate from it.

One method pursued by Laplace to minimize the uncertainty in the theoretical calculation was to try to nail down the masses of the major celestial bodies in the solar system (the sun, the planets and their main moons) and to reduce the uncertainty in these values as much as possible. One illustrative success he achieved was with the mass of Saturn. First Laplace developed a basic tenet of Bayesian probability, now known as Bayes' Theorem, which relates the probability of a theory being true given some new data to both the support it has from previously existing data and from the new experimental data under consideration. He then specified the existing data relating to the mass of Saturn, including the facts that Saturn has sufficient mass to keep its rings from flying off and that it has insufficient mass to perturb the orbits of the inner planets beyond what is observed. The new data he introduced concerned a series of measurements on the mutual perturbations in the orbits of Saturn and Jupiter.

Putting these results into Bayes' Theorem he was able to quantify the level of support that the data rationally bestowed upon his theoretical prediction for the mass of Saturn. He calculated that the data shows there are less than a 1 in 11,000 chance that Saturn's mass deviates from 0.000284738 solar masses by more than 1%. During the subsequent 150 years since this theoretical claim, the accuracy of measuring devices including orbiting telescopes and atomic clocks have increased by orders of magnitude, and yet the current best estimate of Saturn's mass remains well within the narrow range of Laplace's theoretical prediction (Jaynes 1986).

Laplace was the most productive researcher in the history of celestial mechanics and we owe this productivity, in part, to his understanding and development of probability. Before investigating an area he would routinely use probability to calculate the extent to which the existing data supported accepted theory. Only when the data suggested problems with the theory would he throw himself into that area of research. By screening the fit between empirical data and existing theories he could identify those areas where his efforts could be productive in developing either new theoretical understandings that were more explanatory of the data, or new data that could decide between the theoretical alternatives.

The Darwinian nature of Science

Explanations of the evolution of science itself have been framed in terms of a Darwinian processes by prominent philosophers of science of the past century including Karl Popper (1972) and Donald Campbell (1974). The field of Evolutionary Epistemology presents the evolution of systems of knowledge, including science, in terms of a Darwinian process.

In this light the evolution of science might be understood in terms of copies of existing theories being made in scientists' minds, often with variations from the original. Those variations that experience preferential survival are those that are best supported by the data.

Of course the precise manner by which the measure 'best supported by the data' should be calculated is given by Bayesian methods. Indeed E.T. Jaynes dubbed Bayesian Probability 'the logic of science' (2003). The interpretation of this logic of science as a Darwinian process is supported if we consider the probabilistic weighting assigned to each member of a group of competing theories on the basis of existing evidence as a Darwinian selection criteria. Those variants supported by the evidence survive.

Karl Friston, of the Bayesian Brain school of neuroscience, has proposed a theory of mind which models aspects of mental processes as near-optimal Bayesian mechanisms updating mental models from experience (2007). He notes the similarities of this model with that of the model for science:

> *Our capacity to construct conceptual and mathematical models is central to scientific explanations of the world around us. Neuroscience is unique because it entails models of this model making procedure itself. There is something quite remarkable about the fact that our inferences about the world, both perceptual and scientific, can be applied to the very process of making those inferences: Many people now regard the brain as an inference machine that conforms to the same principles that govern the interrogation of scientific data.*

As I have argued above, science may be understood as a cultural re-discovery of the same mechanism used by nature to construct the other numerous knowledge entities which have emerged over the course of evolutionary time. Thus we can conclude that science fits nicely with the four criteria of an adaptive system which we have developed:

The evolution of science according to the operation of a three step 'copy with selective retention' algorithm is central to several of the foremost theories within the philosophy of science (D. Campbell 1965) (Popper 1972).

If we consider the technology of scientific experiments as knowledge entities then we can view the pertinent hypotheses and models of science as internal models which coordinate and predict the relationship of these entities with their environments. These internal models are constructed to minimize the surprise that occurs during operation of the entities. In other words theories are constructed to agree with experimental evidence. Maintenance of the internal model's accuracy through Bayesian updating is implied by the fact that Bayesian probability is the logic of science.

We may consider much of modern technology and engineering as cultural adaptations, based on scientific understanding. They function, in accordance with the principle of maximum entropy to maintain culture in a low entropy state.

Thus we find that our understanding of science itself meets the criteria of a candidate theory within Universal Darwinism. In this light science is but a cultural rediscovery of the methods nature has always used for the construction of knowledge.

9: Universal Darwinism

Views of Universal Darwinism

It is a fact that there are numerous peer reviewed scientific theories which employ Darwinian processes to explain the creation and evolution of their subject matter and it is also a fact that these theories exist across a broad range of science. The mere noting of these facts is the justification for the Universal Darwinism meta-theory.

Previous chapters have reviewed several of these theories and have argued that they share several additional characteristics other than the operation of Darwinian processes. These shared characteristics appear to be implied when considering Darwinian processes in the context of knowledge mechanisms, but they may well provide additional uniformity and depth to the theory. Hopefully the overview and arguments I have offered concerning these characteristics will, if found persuasive, add to the subject matter of Universal Darwinism.

The picture which emerges from this nascent theory provides a unified outline of the evolution of the universe and the structures which have emerged in it. This understanding starts with a network of quantum systems which are fundamental both in the sense that they existed previous to any other structures and in that they form the building blocks of all successive emergent structures. Quantum systems, although fundamental to the web of reality, may be emergent structures in their own right.

Quantum systems are complex information entities. They contain inner models of their environments and these inner models are updated through the experience of the quantum systems in their environments. We are, however, ignorant of the substrate from which quantum systems may have emerged and must be content to limit our exploration of emergent systems to those built with quantum building blocks. We do understand however that quantum processes are the sole means by which information is brought into our web of reality (Lloyd 2007). The other information systems we have examined merely transfer and process this information.

Quantum Darwinism provides a theoretical framework which explains the evolution of quantum systems into the subject matter of atomic physics and chemistry. It does this through encapsulating quantum systems within the framework of adaptive systems and views the evolution into atomic physics and chemistry as a process typical of the evolutionary tendencies of adaptive systems.

When chemistry evolved into life a new adaptive system, with its own emergent laws, was introduced into the universe. Life however also evolved through the operation of a Darwinian process and displays the evolutionary tendencies of an adaptive system.

When nervous systems evolved, a new level of complexity emerged which related living systems to their environments. Leading scientific theories describe the brain as having the properties of an adaptive system and as using a Darwinian process to evolve its model of the environment.

When mental life evolved into culture, complex structures evolved which mediated the relationship between human communities and their environments. The evolution of these structures has perhaps been best explained, within the scientific literature, through employing the concepts of Darwinian processes and adaptive systems.

Thus an understanding of Universal Darwinism provides understanding of the broad strokes of evolutionary history as well as providing the ability to penetrate each level of complexity to significant depths.

A second view of Universal Darwinism might focus on the nested system of constraints which typify complex entities. Murphy's Law that whatever 'can happen' will happen seems to be fundamental to quantum theory. Indeed Alan Guth, one of the founders of modern cosmology, has written (2004):

> In an eternally inflating universe, anything that can happen will happen; in fact, it will happen an infinite number of times.

We might thus attempt to understand what does happen in terms of those constraints which separate what can happen from what cannot happen.

The most fundamental constraint may be that imposed by mathematics; whatever 'can happen' must conform to mathematics. This constraint removes an infinite number of possibilities.

Even in infinite cosmologies, such as the one described by Guth, some phenomena may be much more common than others. For instance although both the set of all integers and the set of integers ending in fifteen zeros are infinite, if we are exposed to a random integer every second of our life we should expect to never encounter one ending in fifteen zeros; even though they are infinite numbers of them they are still comparatively rare.

Within the realm of what can happen the second law tells us that the most frequently occurring phenomena will be those which can happen in the greatest number of ways.

Further, the principle of Maximum Entropy introduces the constraints of scientific law and states that the most frequently occurring phenomena will be those that can happen in the greatest number of ways allowed by the constraints of scientific law.

It might be fruitful to pause for a moment and consider the nature of scientific laws. If we acknowledge that scientific laws cannot exist before the existence of their subject matter we are left to conclude that scientific law must evolve along with their subject matter. The scientific laws governing atomic physics could not have come into existence before there were atoms nor could the scientific laws governing chemistry or biology have been in existence before those structures evolved.

It is apparent that information is fundamental to the nature of the complex entities which have emerged. The deck is stacked against the existence of complex entities. As Richard Dawkins (1986) said:

> However many ways there may be of being alive, it is certain that there are vastly more ways of being dead.

In other words complex entities can only exist if there are scientific laws which constrain what can happen to those outcomes which include their existence. The thesis developed here is that scientific law evolves through the operation of Darwinian processes, providing constraints which allow the existence of specific complex entities. Darwinian processes employed by complex entities conduct searches through design space seeking designs which have the ability to constrain outcomes to those which favour their own existence and survival. These successful constraints are physical structures forming the adaptations of adaptive systems. In this model scientific laws should be seen as the general and common design features of adaptive systems.

Constraints or adaptations are viewed within Universal Darwinism as information-processing mechanisms designed to minimize the system's surprise when encountering its environment, in other words to reduce the error experienced when predictions generated by the system's internal model are tested in the real world. Another way of stating this is that an adaptive system expects to survive and its adaptations are designed to make this expectation a reality. Again Universal Darwinism may provide a unified understanding having explanatory power over a number of layers of emergent reality.

The last view of Universal Darwinism I will discuss focuses on Universal Darwinism as the path of knowledge. For our purposes knowledge has been defined as the ability of models to make accurate predictions of what they will encounter. In previous chapters we have explored the sense in which such models exist in the form of the wave function of quantum systems, the genome of biology, the pattern of neural connections of the brain and the literature of science.

We have described physical reality, at its most basic level, as composed of a quantum web of information exchange. Central to the definition of information is expectation. Of all the configurations or designs which may be assumed by information systems, those that can survive and have a physical existence share two characteristics:

1) A design which 'expects' to survive. That is, one which will survive given the situation and resources it 'expects' or is designed to encounter.

2) Its expectations are realistic and accurate. That is, they are based on knowledge.

Knowledge or model accuracy does not happen by accident, it must be caused, and mathematically there can be only one cause: inference. Inference must be achieved in a manner consistent with Bayes' theorem. I have argued that the Darwinian process is a means of implementing Bayes' theorem in physical reality and that wherever knowledge is found in nature there are scientific theories which explain its creation and evolution as the product of a Darwinian process. Thus Universal Darwinism provides us with a meta-theory concerning knowledge entities.

An implementation of knowledge within a physical entity implies complexity and the physical survival of a complex entity is dependent on knowledge. The specific type of knowledge required is how the entity can interact with the environment so that its low entropy state can be maintained in exchange for increased environmental entropy. This interplay between complexity and knowledge requires the production of adaptations within the entity to optimize environmental encounters. This increases complexity and this increased complexity in turn requires greater knowledge to sustain it.

The physical implementation of such an entity unleashes a never-ending search for both greater knowledge and greater physical complexity. Universal Darwinism may be seen as an attempt to explain those complex and scientifically interesting entities which have emerged in our universe over evolutionary time as products of this path of knowledge.

The most fundamental layer of reality we have detected, the one at the beginning of this path of knowledge, is that of quantum systems. I suspect this may only be due to our current lack of sophistication in detection and that more fundamental layers may eventually be found. Nevertheless the quantum reality predates all other layers of emergent reality we have identified and forms a closed system or web of reality.

We may view the evolution of the quantum constituents of sub-atomic particles into atoms and more complex chemistry in terms of Quantum Darwinism. This Darwinian process allows a quantum system to probe its environment searching for and selecting the optimal low entropy states from all those available, thus allowing greater complexity to be discovered and survive.

The evolution from chemistry to life is characterized by the control of chemical laws by biological ones; a new knowledge model (the genome) capable of orchestrating cellular chemistry evolved. To support the added complexity, entailing sophisticated constraints on cellular chemistry, greater knowledge was required to optimize interactions with the environment. This interplay of greater complexity requiring increased knowledge, which in turn made possible even greater complexity, is the story of the rapid evolution of life on earth.

When nervous systems evolved they formed a new knowledge model. Features of the external world were modelled within the nervous systems of organisms. The knowledge contained within this new model allowed more refined interactions with the environment but also added greater complexity to the organism, which in turn required greater knowledge in order to sustain itself.

Even some relatively primitive organisms have the capacity for rudimentary learning, but the sophistication of models which they could learn tended to increase along with the sophistication of nervous systems over evolutionary time. The human brain has the ability to learn and imitate hugely complex models. This involves the individual in making copies of models available within his or her culture and a process of cultural evolution often described within the scientific literature in terms of a Darwinian process.

The power and complexity of cultural models seems to have followed an exponential trajectory over the course of human history. Such models today take many forms including business plans, engineering diagrams, text books and the scientific literature. The huge array of cultural constructions we have built, which usually mediate the manner in which we experience our environment, are often reliant on some form of model even if it is only an idea in an individual's mind. The low entropy nature of our cultural constructions is maintained through the consumption of huge amounts of energy, often in the form of fossil fuels, which serves to swap entropy from our culture to the environment.

In this manner human culture has gained the ability to orchestrate many of the phenomena of other emergent levels of reality, including psychology, biology, chemistry and physics. Utilizing the knowledge of our models we are able to constrain these phenomena to serve our own purposes. As our knowledge expands so does our control over natural processes, leading to a more complex culture which in turn requires greater knowledge to sustain. In this manner culture continues to blaze the path of knowledge.

This view of Universal Darwinism provides us with a unified and integrated view of the evolution of information processes and knowledge entities. Although this view is scientific it does acknowledge that humans have a special place, as growing tips, on this ageless, sprouting tree of knowledge.

10: The Worldview of Universal Darwinism

The comprehensive nature of Universal Darwinism may inform our worldview.

In this connection the philosopher Daniel Dennett has observed that the theory of natural selection is so important, and to some so threatening, precisely because its acceptance radically alters our worldview; it provides scientific explanations for some of our 'big questions' concerning life, meaning and purpose (1995).

> *In a single stroke, the idea of evolution by natural selection unifies the realm of life, meaning, and purpose with the realm of space and time, cause and effect, mechanism and physical law.*

By extending the principles of natural selection Universal Darwinism offers an even more radical updating of our worldview and provides answers to even deeper 'big questions'.

Why there is something rather than nothing has been called "philosophy's central, and most perplexing, question" (Rundle 2004). Universal Darwinism offers an answer. If we accept that the 'something' refers to the existence of complex entities then Universal Darwinism offers a rather complete explanation; complex entities are brought into existence through the operations of Darwinian processes which infer knowledge, in support of survival, from information existing in the entity's environment.

However, its power resides not only in answering philosophical conundrums, it also provides insight into personal questions we might struggle with such as 'what are the pros and cons of basing belief on faith?'

Perhaps most importantly Universal Darwinism offers a unified approach and accessibility to much of scientific subject matter. This may make it easier to incorporate scientific understanding into our worldview.

Faith and Universal Darwinism

A group of prominent Darwinists, including Stephen J. Gould, have proposed a detente between religious faith and scientific understanding (Gould 1999). They suggest that while science may provide the appropriate path to knowledge in some areas, faith may be the appropriate method of achieving knowledge or understanding in others such as ethics, morality and spirituality. This accommodative view is not shared by other Darwinists such as Richard Dawkins, who see religious faith as a societal evil promoting an inclination towards warfare and the routine indoctrination of children into cult-like superstition.

Science as seen through the lens of Universal Darwinism may provide a context to better understand this debate and perhaps offer some clarification to it.

The employment of faith as a method of knowledge is not restricted to religion. The first definition of faith offered by an on-line dictionary (Dictionary.com 2010) is:

confidence or trust in a person or thing.

This definition suggests an intuitive basis for faith but does not exclude confidence achievable by other methods such as scientific inference. The second definition offered makes the distinction clear:

belief that is not based on proof.

Faith might well be characterised as confidence in a belief regardless of the proof or evidence. As science is committed to belief based solely on evidence it is difficult to see a middle ground. Stephen J. Gould finds this middle ground through arguing that scientific methodology is applicable to some subject matter while faith is more appropriate to others.

Universal Darwinism's contribution to this debate may result from its understanding that science is but one method of accumulating knowledge out of the many that have evolved during the history of the universe. These knowledge entities also include quantum systems, genetics and brain functions. At first the acknowledgement of multiple processes for producing knowledge might make us more sceptical that any specific form such as science might offer a uniquely correct method of arriving at knowledge.

On some subjects science does not even contain the most highly evolved forms of knowledge. For instance the introduction of antibiotics into the realm of science was accomplished by observing and copying the genetic knowledge of bacterial warfare which existing organisms had discovered and accumulated over eons. Antibiotics were designed and produced by genetics and only adopted by science. However Universal Darwinism informs us that all known knowledge entities, including Science, accumulate their knowledge through the process of inference, which is the rational construction of knowledge from evidence.

A common definition of Universal Darwinism is the collection of scientific theories that utilize the Darwinian process to explain the creation and evolution of their subject matter. Given that such theories span the realm of scientific subject matter and that they have no scientific rivals, we are left with the suggestion that Darwinian processes may be the only method of knowledge accumulation operating in the universe. It is this suggestion which directly challenges the notion that faith may be a valid method for obtaining correct beliefs.

The argument for the uniqueness of Darwinian processes as knowledge mechanisms may require some explanation. This argument is most directly supported through the link established between Bayesian probability and Darwinian processes (J. Campbell 2009). Most beliefs lack certainty. To be certain in ones beliefs may be psychologically appealing but is extremely unlikely to be justified. To justify certainty in a particular belief one would have to be certain that the understanding that underlies it will never improve at any time in the future. It also means that all other people holding incompatible beliefs are certainly wrong. Certainty in belief is not usually a defensible position, instead we deal with degrees of plausibility or probability. It has been proven that the unique mathematical method for extending logic from true/false types of statements to those concerning degrees of plausibility is Bayesian probability (Kirby 2003).

Specifically Bayesian probability contains theorems for calculating the probability that competing beliefs or hypotheses are true. These theorems require that in order for our beliefs to be optimal they must be updated whenever we receive pertinent new data. While these theorems form the basis for deciding between competing scientific hypothesis we must remember that they are mathematically true and apply to anything conforming to the rules of mathematics which is, as far as we know, everything. Thus it may be argued that anything capable of accumulating knowledge must rely on data or evidence. As it has been argued that Bayesian updating is isomorphic with Darwinian selection (J. Campbell 2009), we have a strong argument that Darwinian processes are but physical implementations of the mathematical rules governing any form of knowledge accumulation.

Thus beliefs founded on data, evidence or proofs, as opposed to those based on faith, are much more likely to be true.

Still much of this may seem very counterintuitive. We receive many if not most of our judgements from our intuition rather than through the laborious process of carefully weighing all the data in our conscious mind. Many of these unconscious judgements or beliefs seem to serve us well, causing us to flee from danger when necessary and to take appropriate actions in a variety of circumstances. In short many intuitive beliefs are justified.

The answer to this conundrum may be that although the formation of intuitive beliefs does not involve conscious selection from data, they are themselves formed through selection from data by other unconscious mental or genetic mechanisms. The Bayesian brain theory, a comprehensive, general theory of mental mechanisms, argues that the mental models or beliefs employed by our brains are based on inference from sensory evidence or data (Friston and Klass 2007). These mechanisms for forming unconscious mental models have been honed by natural selection over eons to provide accuracy in those circumstances where survival or reproductive success is at stake. This is a crucial point. It implies that those unconscious mechanisms may give us reliably accurate guidance in circumstances such as predator avoidance which have a long evolutionary history (long before humans existed), but may not be so reliable when dealing with circumstances having a shorter evolutionary history or ones involving little consequences for their host's survival.

As our intuition is beneath our conscious radar we are little aware of its operation but obviously in many circumstances our unconscious mind provides our conscious mind with judgements. Most of these are mundane and automatic like how to move our leg when stepping over an object. For these we usually get extremely accurate answers, we rarely put a foot wrong. This accuracy may well be attributed to the long evolutionary history we have inherited dealing with terrestrial locomotion. We may well have conscious faith in these unconscious judgements. Although we have little conscious evidence or data on which to base them we may be assured by the fact that these judgements represent knowledge based on evidence accumulated over evolutionary time.

Unfortunately the situation is much different when our intuition provides judgements concerning issues that are of little evolutionary consequences, judgements whose accuracy has not had the benefit of being honed by evolutionary processes. Our unconscious mind seems to have little hesitation in providing answers in these circumstances but the accuracy of those answers is highly suspect.

Pascal Boyer, in his book Religion Explained (2001), argues that it is such untested intuitive models which form religious beliefs. He contends that the hugely complex, often accurate judgements we make about social situations are inappropriately applied to wider questions. If for instance, the fellow next to us picks up a stick we are usually able to come to a fairly accurate snap judgement on his intentions and our optimal response to his actions. However when we are confronted with some mysterious natural process, perhaps thunder and lightning, we rely on these same unconscious mechanisms to provide answers. Often these answers tend to be highly anthropomorphic such as suggesting that someone up in the sky may be angry with us and is giving us a warning.

It is this type of intuitive judgement of which we should be highly sceptical because it is based on little underlying data. Such scepticism is difficult for us for it means the conscious mind has to overrule intuition. Michael Shermer, editor of the Sceptical Inquirer and author of <u>Why People Believe Weird Things</u>, claims people often have weird beliefs because we tend to receive them as intuitions and then use our rational mind to justify them rather than to overrule them.

These types of beliefs do not justify our faith. They are not based or evidence either conscious or unconscious and are therefore highly implausible.

Universal Darwinism may offer further insights even into the circumstances where faith in myths is perhaps appropriate. It makes clear that our consciousness has been crafted over evolutionary time to focus on issues which promote our survival and reproductive success. It is little wonder that thoughts of sex, resources and status are never far from our minds. Yet our minds also have an interest in the bigger picture, the context in which we exist. Theories of mind such as the Bayesian brain propose that the primary task of our brains is to model the world around us, thereby allowing us to conduct optimal outcomes for ourselves through our interactions with it. Any successful modeling of the outside world soon reveals that it is not all about us. It has its own concerns and marches to the beat of its own drummer. It is overwhelmingly vast especially in comparison with our self. Our ignorance of it is likewise vast and yet knowledge of it is crucial for survival. Our ignorance in light of its enormity, objectivity and importance lead to feelings of awe and humility and desperation for insight into the nature of this unknowable reality and our relationship to it.

Every culture known throughout the course of time has had a religion, serving to provide a comforting tale of the big picture and the part played by humans in it. Usually these contain a creation myth defining a group of people through their special creation overseen by some supernatural force or process.

Just as individuals are dependent upon the world around them for survival they are also dependent on their community and the cohesion of their group. Religion and shared myths can strengthen this cohesion and confer a selective advantage to those groups which most strongly bind their members. Under these circumstances the literal accuracy of these myths is of little importance. In any case, before the advent of science humanity possessed no powerful systematic method of accumulating knowledge, little data was available to earlier cultures concerning the nature of the big picture, so in a sense each was equally plausible. Darwinian success of the culture depended to an extent on its ability to cohesively bind its members into common undertakings and it was in this context that religions could assist. Much of this binding took the form of belief in common myths, in other words in beliefs not based on evidence or proof. In this context faith was a virtue.

When Nietzsche rocked the world in the 1880s with his proclamation that 'God is dead', the intellectual world had been well prepared to resonate with his message having had twenty years digesting Darwin's work. Darwin and others had produced a plethora of evidence for the theory of natural selection. Darwin himself conducted painstaking studies of barnacles, orchids, earthworms and coral reefs, each a definitive work, and each documenting how the theory of natural selection explained the many features of the subject matter under consideration. In The Origin of Species he brings decisive evidence to bear with subject matter as diverse as artificial selection and the geological record. Overwhelming evidence was presented for a theory which explained our relation to all other living things and the details of our creation. Since Darwin the supporting data has continued to build and new sources of supporting evidence have been discovered, especially that based on DNA studies.

The most powerful argument for the existence of God has long been 'the argument from design'. This is based on the reasoning that the existence of the complex knowledge entities found in nature could only have been created by an entity much more knowledgeable and powerful than the created entity. Darwin instead proposed a simple, bottom up, method of accumulating knowledge, the Darwinian process, which is natural and does not require a complex and powerful creator. The evidence suggests that Darwin's explanation is the correct one.

Since the time of Darwin society has also evolved. While religion was once a defining characteristic of a society and membership in it was often mandatory, today most advanced societies specifically exclude religion from being a compulsory criterion for membership. Cohesion and cooperative actions are orchestrated by other mechanisms such as market forces, sports teams, nationalism and legal systems. Faith-based creation myths no longer provide much societal benefit, at least in the sense of conferring survival.

Prior to Darwin faith in myths concerning the bigger picture may have been the best one could do, both for oneself and for society; now such faith has not only outlived its usefulness on a social level it also amounts to wilful ignorance regarding factual matters on a personal level.

Miracles of Universal Darwinism

Three roads to a religious experience

The central religious message is 'This is not about you.' The great religions that burst on the scene starting around 3,000 BC are unanimous on this point: there is a God or Gods that 'this' is 'about'. We are merely God's creation or His dream or emanate from Him in some other way. This central message makes way for the spiritual experience of awe and humility in the face of a vast creative power.

I've always had a soft spot for the religious experience. To be made suddenly aware of a cosmic truth overwhelms us with feelings of wholeness, awe and well being. Strange as it may seem, science also provides intense religious experiences to those who follow it. Einstein affirmed '*I assert that the cosmic religious experience is the strongest and the noblest driving force behind scientific research.*' (Simpson 2004)

Science, because of its deep connection with truth, may even provide a superior religious experience for those exposed to its mysteries. To explore this idea, we will examine three versions of a single mystic insight. One of these versions is scientific.

I first encountered this religious story in the form of a Brahmin parable that went something like this: 'In the beginning there was God and only God and he was a unified entity. He grew bored with being alone and contrived to split himself into an infinite number of separate pieces that would slowly come to know their situation, find their way back together and gradually become aware of their divinity. This process composes the essential history of the universe.'

George Spencer-Brown, the Oxford logician and western mystic, describes another version of this same mystery (Spencer-Brown 1979):

> *Let us consider, for a moment, the world as described by the physicist. It consists of a number of fundamental particles . . . Now the physicist himself, who describes all this, is, in his own account, himself constructed of it... Thus we cannot escape the fact that the world we know is constructed in order (and thus in such a way as to be able) to see itself... This is indeed amazing. It seems hard to find an acceptable answer to the question of how or why the world conceives a desire, and discovers an ability, to see itself, and appears to suffer the process. That it does so is sometimes called the original mystery.*

Universal Darwinism presents this same mystery in scientific terms: 'In the beginning, at the time of the Big Bang, the universe was a single integrated point. It then dispersed itself into simple and isolated forms. However, over billions of years starting from the Big Bang evolutionary processes began to create progressive order and integration from the parts. Physical processes evolved ever greater complexity eventually producing on earth the complex chemistry necessary for life. With the evolution of living creatures, each generation contained many more variable individuals then could be supported by their environment. Only those individuals with the variations, or adaptations, most suitable to their environment survived and in turn produced offspring. Successful adaptations are knowledge of the environment and how to make a living in it. More powerful adaptations or ways of knowing evolved, including human scientific knowledge. In this manner living things gradually came to know the universe of which they are a part and thus the universe came to know of itself.'

These three presentations of essentially the same mystery can be compared in the impact of the religious experience they induce for those open to them. How do they inform us and lead us to a true and awed understanding of the situation in which we find ourselves? First let us consider the Brahmin tale. This parable adequately sums up our situation as stuff of the universe having knowledge of the universe. Confrontation with this awesome fact leaves us with a feeling of marvel and wonder. But it ends there. There are no supporting details. If we insist on details of the processes by which pieces of the universe come to know their wholeness and divinity, we are left only with a mystery. The religious experience is intense but it is not able to sustain any independent probing for details.

Spencer-Brown's account is presented in more modern terms. In fact the work from which his quote was taken is a breakthrough in mathematical logic. In this work he shows that the mere drawing of a distinction, or the act of splitting apart, provides a sufficient basis from which to derive all mathematical logic. In other words once there is in any sense diversity, all of mathematics is implied. As he notes in this work, mathematics is a tool by which we come to know ourselves. His account presents us with the same mystery as the Brahmin parable but has an added richness in the mathematics he explores. Unfortunately if we press awkward question as to the details of the universe coming to know itself, as Spence-Brown says, it is a mystery.

Universal Darwinism's account presents exactly the same mystery as the other two. If in the Brahmin parable you substitute the words 'The Universe' for 'God' and 'It' for 'He', the result will serve as a scientific version of the mystery. We can equally savour with awe the miraculous nature of our situation from this version. We are part of a universe that is coming to know itself. However with science we can probe for more details and be richly rewarded. Science provides surprising details of adaptations and the knowledge they provide of the world. It provides details of the different forms of self knowledge that the universe has evolved: genetic knowledge, instincts, learned behaviour and higher intellectual knowledge. Truly there are limits to this knowledge. Science is open about its ignorance, but the boundaries separating knowledge from ignorance are not static in science as they are in religion. Science is involved in a rapid expansion of knowledge.

Religion teaches us that 'this' is not about us, that we should be humble and in awe of the divine reality that 'this' is about. Universal Darwinism agrees with this view at least to the extent that we are but bit players in a vast and awe inspiring cosmic drama, but it also acknowledges another reality, the reality of our role as bit players where we are concerned with personal things such as survival, reproduction and status. Some religions teach that this aspect of our lives should be shunned, that the road to spiritual enlightenment consists of reducing the role of self or ego. Universal Darwinism explains the self and our concern with it as part of the bigger picture. Each generation in this cosmic drama has a kind of duty to probe the possibilities of their situation and attempt to survive and prosper.

Universal Darwinism also outlines the dangers of an over large focus on the self. Memetics explains the self as a meme complex that resonates with our minor role and is concerned with the details composing our individual dramas, largely survival, sex and the acquisition of resources and status. If left unchecked this memeplex may hijack all our mental machinery to concerns of the self. The danger of an unconstrained self is that it tricks us into thinking of ourselves as all important, as the only reality we should be concerned with. In short it traps us into identifying ourselves as mortal. Universal Darwinism shows us that this danger may be mitigated by meditation or science. Both of these activities illuminate the self as a bit player in the cosmic drama and while it is our karma to play out this role it should not blind us to the greater reality.

Einstein was clear on this point as he was on so many others. He saw science as providing us with the context of a cosmic religious experience where the self is constrained to its proper proportions (1930).

> *A human being is a part of a whole, called by us universe, a part limited in time and space. He experiences himself, his thoughts and feelings as something separated from the rest...a kind of*

optical delusion of his consciousness. This delusion is a kind of prison for us, restricting us to our personal desires and to affection for a few persons nearest to us. Our task must be to free ourselves from this prison by widening our circle of compassion to embrace all living creatures and the whole of nature in its beauty.

Einstein and the evolution of religion

Not only was Einstein one of the greatest scientists of all times but he also made significant contributions to philosophy including the philosophy of religion. He often used religious metaphors in his scientific discussions such as his famous statement that God does not play dice but the God he referred to was not an anthropomorphic God, rather it is the God of Spinoza, that is God as natural law. Many have deluded themselves that Einstein shared their belief in a personal God. Einstein was pestered by ardent religious followers from many sects. Recently a 1954 letter of his replying to a Jewish devotee was discovered in which Einstein's views are made crystal clear (1954):

> *For me the Jewish religion like all other religions is an incarnation of the most childish superstitions.*

Einstein believed that religion, as a component of human culture was subject to evolution. He outlined this thinking most clearly in an article written for the New York Times Magazine in 1930. This article outlines religion as having evolved through three stages. Just as in biology where primitive forms such as blue-green algae still exist in abundance the first two stages of religious evolution outlined by Einstein still have large numbers of adherents.

Roughly he outlined the three stages as (1930):

1) Religion based on fear.

With primitive man it is above all fear that evokes religious notions - fear of hunger, wild beasts, sickness, death. Since at this stage of existence understanding of causal connections is usually poorly developed, the human mind creates illusory beings more or less analogous to itself on whose wills and actions these fearful happenings depend. Thus, one tries to secure the favor of these beings by carrying out actions and offering sacrifices which, according to the tradition handed down from generation to generation, propitiate them or make them well disposed towards a mortal. In this sense, I am speaking of a religion of fear. This, though not created, is in an important degree stabilized by the formation of a special priestly caste which sets itself up as a mediator between the people and the beings they fear, and erects a hegemony on this basis. In many cases, a leader or ruler or a privileged class whose position rests on other factors combines priestly functions with its secular authority in order to make the latter more secure; or the political rulers and the priestly caste make common cause in their own interests.

2) Religion based on morals.

The social impulses are another source of the crystallization of religion. Fathers and mothers and the leaders of larger human communities are mortal and fallible. The desire for guidance, love, and support prompts men to form the social or moral conception of God. This is the God of Providence, Who protects, disposes, rewards, and punishes; the God who, according to the limits of the believer's outlook, loves and cherishes the life of the tribe or of the human race, or even of life itself; the comforter in sorrow and unsatisfied longing; he who preserves the souls of the dead. This is the social or moral conception of God. The Jewish scriptures admirably illustrate the development from the religion of fear to moral religion -- a development continued in the New Testament. The religions of all civilized peoples, especially the peoples of the Orient, are primarily moral religions. The development from a religion of fear to moral religion is a great step in peoples' lives. And yet, that primitive religions are based entirely on fear and the religions of civilized peoples purely on morality is a prejudice against which we must be on our guard. The truth is that all religions are a varying blend of both types, with this differentiation: on the higher levels of social life, the religion of morality predominates.

3) Religion based on the cosmic religious experience.

In general, only individuals of exceptional endowments and exceptionally high-minded communities, rise to any considerable extent above this level. But there is a third stage of religious experience which belongs to all of them, even though it is rarely found in a pure form: I shall call it cosmic religious feeling. It is very difficult to elucidate this feeling to anyone who is entirely without it, especially as there is no anthropomorphic conception of God corresponding to it.

The individual feels the futility of human desires and aims at the sublimity and marvelous order which reveal themselves both in nature and in the world of thought. Individual existence impresses him as a sort of prison and he wants to experience the universe as a single significant whole. The beginnings of cosmic religious feeling already appear at an early stage of development, e.g., in many of the Psalms of David and in some of the Prophets. Buddhism, as we have learned especially from the wonderful writings of Schopenhauer, contains a much stronger element of this.

The religious geniuses of all ages have been distinguished by this kind of religious feeling, which knows no dogma and no God conceived in man's image; so that there can be no church whose central teachings are based on it. Hence, it is precisely among the heretics of every age that we find men who were filled with this highest kind of religious feeling and were in many cases regarded by their contemporaries as atheists, sometimes also as saints. Looked at in this light, men like Democritus, Francis of Assisi and Spinoza are closely akin to one another.

How can the cosmic religious feeling be communicated from one person to another, if it can give rise to no definite notion of a God and no theology? In my view, it is the most important function of art and science to awaken this feeling and to keep it alive in those who are receptive to it.

Given Einstein's immense stature and his inspiring beliefs, his religious views have had remarkably little impact. Of the tens of thousands of religious and quasi religious cults that seem to have sprouted up like mushrooms there does not appear to be a cult of Einstein, unless of course it is within science itself.

Other than the odd philosopher, seer and saint sprinkled throughout the centuries of history, those that Einstein identified as sharing the cosmic religious experience are largely scientists (Einstein, Science and Religion 1930).

> *I maintain that the cosmic religious feeling is the strongest and noblest motive for scientific research. Only those who realize the immense efforts and, above all, the devotion without which pioneer work in theoretical science cannot be achieved are able to grasp the strength of the emotion out of which alone such work, remote as it is from the immediate realities of life, can issue. What a deep conviction of the rationality of the universe and what a yearning to understand. Were it but a feeble reflection of the mind revealed in this world, Kepler and Newton must have had to enable themselves to spend years of solitary labor in disentangling the principles of celestial mechanics! Those whose acquaintance with scientific research is derived chiefly from its practical results easily develop a completely false notion of the mentality of the men who, surrounded by a skeptical world, have shown the way to kindred spirits scattered wide through the world and through the centuries. Only one who has devoted his life to similar ends can have a vivid realization of what has inspired these men and given them the strength to remain true to their purpose in spite of countless failures. It is the cosmic religious feeling that gives a man such*

strength. A contemporary has said, not unjustly, that in this materialistic age of ours the serious scientific workers are the only profoundly religious people.

Still we might ask why this vision of religious feeling is not more widely embraced outside of the scientific community. The short answer may be that science is hard to understand and one must devote them self wholeheartedly to a huge effort of study in order to understand even a small subset of science.

Unlike many conventional religions where one need only decide to take a leap of faith, science requires a commitment to understand reality on the basis of the available evidence. This understanding and the explanations of science based on it have been developed in a largely ad hoc manner where each small piece of subject matter comes with its own idiosyncratic jargon, methods and orientation and requires years of solitary labour to master.

Universal Darwinism, through offering a simple and consistent explanation across much of scientific subject matter, may make Einstein's cosmic religious experience accessible to greater numbers and serve to 'awaken this feeling and to keep it alive in those who are receptive to it'.

Who's got the miracles?

Miriam Webster provides a couple of definitions for 'miracle'.

1) *an extraordinary event manifesting divine intervention in human affairs*

2) an extremely outstanding or unusual event, thing, or accomplishment

The main difference between these two definitions is the reliance of the first on the divine, on supernatural agents while the second leaves the cause of miracles unspecified.

Miracles produce the religious experience by confronting us with amazing phenomena often extremely difficult to explain. Many find such ignorance difficult to accept and are more comfortable explaining miracles as due to supernatural agents. The only problem is that these 'supernatural' agents, whether they are the Holy Ghost, God or Angels, are almost by definition incapable of detection and it is therefore impossible to verify that the said agent was indeed responsible for the miracle. For instance if a prayer to God is said over a patient with 'incurable' cancer, and that patient recovers, how can it possibly be verified that the healing was done by God when God is not detectable and the means by which he conducts the cure unknown? After all a certain number of people with 'incurable' diseases do recover without prayer and many die who receive prayers. How can attribution of healing to God be anything more than wishful thinking taken on faith? Indeed the record is replete with documented cases of faith healers wilfully conning those gullible enough to believe in this kind of miracle.

It is also a routine, well documented fact that many people with otherwise 'incurable' conditions recover after receiving medical treatment. This, of course, is not commonly thought of as a miracle because the cure is not attributed to supernatural agents. In order to be approved for use a new drug or surgical procedure must pass a battery of experimental tests to prove that it is effective in treating the condition. The tests must be verifiable by others and produce highly reliable results. Almost always the effectiveness of the medication or surgery is readily explained by pre-existing medical knowledge. In a sense it is no miracle but it may often be considered an 'extremely outstanding or unusual event, thing, or accomplishment', especially when it is newly discovered.

That medical knowledge is able to reliably cure so many lethal conditions is a phenomenon truly deserving of awe and is often described as a miracle. We can attribute cures that occur for unknown reasons to supernatural agents but aren't we only fabricating a fiction to mask our ignorance?

Throughout history, religious or supernatural means have often been invoked in the quest for power. Magical spells have been cast and God has been implored to give us superhuman powers or to destroy enemies. Sometimes he is said to have complied as in the Biblical tale where the Egyptians were engulfed by the Red Sea. Those occasions on which the supernatural is thought to have lent us their powers are occasions that strike us with awe and cause us to reverently give thanks. They are miracles. And yet this means of invoking power has been highly unreliable. Indeed the evidence suggests that none of these myths of God exerting his power on our side have any factual truth, that all our uncountable invocations have never made the slightest difference except perhaps in the psychology of those performing them.

The only reliable method we have ever had for invoking power has been knowledge and by far the most powerful knowledge we posses is science. Using science we can fly further and faster than any bird. In fact we can fly to the moon and there is little doubt that soon we will fly to Mars. Using science we have the power to engulf our enemies in a mini-sun if we lack the wisdom to refrain. But this is not a divine miracle; there are no supernatural agents involved. These phenomenons are fully explained by our scientific knowledge. And yet they are worthy of awe and of a deep and humble reverence for our situation.

The Miracle of Time Travel

Evolution works through the cumulative building of knowledge. Our genetic and intellectual 'knowledge' of the world in which we live provides us with the multitude of adaptations that allow us to survive. Cumulative knowledge has the effect of providing a time warp for the construction of complex entities such as ourselves. This is of great significance for each of us. Starting at conception, each of us has individually grown through the same evolutionary process that life followed during the three and one half billion years taken to achieve the transformation from first life to human. After birth each of us has undergone intellectual growth that is the cumulative result of a hundred thousand years of cultural evolution. Each of us has relived and has been created by a personal journey through the entire expanse of replicator-based evolutionary knowledge.

It took evolution many billions of years to gain the knowledge required to construct life from chemistry. The specialized circumstances required to persist complex chemistry are rare in the universe. A defining characteristic for life is its ability to provide and maintain the circumstances required for the existence of specific complex chemical processes.

Our bodies routinely construct life from chemistry in a matter of days. Chemicals, derived from the air and our food, are woven into the complex chemicals required to create new cells. Egg and sperm cells, which are each individual's starting point are created in this manner and can be seen to parallel the evolution of chemistry to life. The knowledge gained by evolution over billions of years is now used to produce new life in days. Once the way through design space is known you can get there quick.

It took evolution about three and a half billion years from when life first evolved to gain the knowledge required to build human beings. It was a long journey encompassing ancestors of single-celled life, fishes and mammals. After sex cells join to form an embryo in the womb it takes only nine months to journey from single celled life to a human being. The embryo's developmental journey parallel's the development of our ancestor's embryos. Their slowly won knowledge is now available to produce a human baby in a cosmic instant. All of us, in embryonic form, underwent this vast journey through design space in only nine months. Once the way through design space is known you can get there quick.

The details of this process reveal that our embryos follow the same evolutionary path as the embryos of our ancestors. For instance, at one point in their development our embryos had the forbearers of gills. Up to this point our embryological development was very similar to that of our fish ancestors; subsequent to this point in development our fish ancestors went on to further develop their gills and other fish characteristics. After this developmental stage our embryos go on to develop the attributes of our more recent air breathing ancestors. In this manner our individual development follows the path taken by the evolution of the ancestors of our species.

There is only one path known for single-celled life to man; it was discovered and mapped over three and a half billion years as a result of evolution's relentless search through design space. The knowledge of how to make this transformation is stored in our genes as a recipe. Every human being is built with this recipe (although it varies slightly with each individual).

It took evolution about one hundred thousand years from first modern man to develop our culture. Cultural attributes, or memes, were at first passed between individuals and generations through mechanisms like imitation and storytelling. More recently storage mechanisms like books and computers have sped up cultural transmissions. The advent of these storage mechanisms has made possible the development of science where the extent of detailed knowledge far exceeds the retention capacity of any single individual.

The evolutionary nature of cultural attributes is discernable in diverse studies. For instance the development of tools such as spear heads has been shown to follow an evolutionary pattern amongst all peoples using them. These weapons developed slowly through time, each development clearly a variation on an existing theme. Their persistent patterns have been used to map the diffusion of human groups throughout the world.

It is clear that our current world culture was inherited with variations from previous states. Evolution took one hundred thousand years to find these paths through design space. Each of us as young people, learn a portion of this world culture. We start with some of its earliest accomplishments: speech, writing and how to make fire. Any cultural area that is learned in detail starts with a historical overview of how the current subject matter came to be.

Although it took one hundred thousand years to evolve, we learn the majority of our cultural inheritance during our first twenty odd years. Once the way through design space is known you can get there quick.

Attaining maturity entails a re-enactment of the evolutionary drama. During development we whiz through vast expanses of evolutionary time and receive their benefits in a cosmic instant. At maturity, as adults, we have reached the end of this journey through time and find our self in the present. There is no longer any untapped repository of evolutionary knowledge that can prompt us through eons of progress in an instant. We have caught up to evolution's production of knowledge, we embody it all, and now the only way forward is to take part in the creation of new knowledge. We are in the present; we are active agents in evolution's great exploration of design space. We are searching for the good designs, those designs with the ability to endure.

This transition from developmental to a mature status is fundamental to an accurate understanding of who we are and our place in the universe. Our developmental inheritance endows us with the full range of evolutionary achievements. We are the product of a vast range of knowledge slowly accumulated by evolution over billions of years and a significant proportion of our life is taken up in the re-creation of this knowledge in a human form. In our mature state we take our inheritance of evolutionary knowledge and venture out into the present to explore design space and to search for new knowledge to further supplement our vast inheritance.

The Miracle of Immortality

Although religions agree that 'this' is not about the self, that there is a much more cosmic reality in which the self is but a bit player, many religious sects (especially those in the Christian and Islamic traditions) make the promise of an everlasting life for the self. Other sects such those in the Buddhist and Hindu traditions claim that what immortality can be achieved can only be achieved to the extent that the self is given up.

Universal Darwinism clearly sides with the latter and provides deep explanations that detail our intimate relationship with the immortal cosmic reality. It explains that it is to the extent that we identify ourselves with this reality that we will actually be immortal.

Many religions and spiritualists promise the possibility of immortal life. This is a deeply comforting promise as fear of death is perhaps our deepest and most basic fear. The Christian religion decrees that there are two types of afterlife (three if we count purgatory). Whether one goes to hell or heaven depends on whether one is a good Christian as defined by the sect making the rules.

There is of course absolutely no evidence for an afterlife of this kind. Every single medium who has subjected themselves to scrutiny has been exposed as a charlatan. No convincing details concerning the inner workings of this afterlife are available.

From the sect's point of view, promise of an afterlife is a wonderful control mechanism enforcing faithfulness amongst their followers. Given the vast number of strict and mutually exclusive sets of criteria for successfully achieving the afterlife expounded by various sects it is logically impossible for any other than the smallest minority to achieve a pleasant afterlife even if some particular version were to be true. Each sect however believes that they are that minority and will live forever if only they follow their teachings. It is easy to see why sects with this item of faith have a competitive advantage.

From the believer's point of view, this is the usual trade off required by faith: give up any doubts and independent thought in return for a comforting story, give up any attempt to make real sense of our situation in return for anaesthetic.

When we ponder what a believer might construe to be the details of their afterlife we quickly encounter problems. Are we the same person in the afterlife that we are in this one? Do we have the same interests, the same imperfections? Are we prone to boredom, anxiety, or irrational fears? If we are vastly different in the afterlife can we remember who we were in this one, is there any continuity of our egos?

My father is 88 years old and a devout Christian. Before he became afflicted with Alzheimer's he expected to have an afterlife where he would be reunited with his deceased daughter and other departed loved ones. He doesn't talk of this now and would not be able to comprehend the question if asked. He is now almost totally unaware of who he is or what his life was. I sometimes tell him the story of his life, details of what he did in his working life, stories of his friends, the adventures he undertook. Sometimes these accounts stir distant memories. I have recently come to understand that there is more of 'him' alive in me then there is in him. When he dies and were he to enter the afterlife in his present state and be reunited with my sister he would not recognize or remember her. Would he be restored to some state earlier in his life? Would he be the same person at all?

It soon becomes clear that it is pointless to ponder the details. You are free to make up any story you want, any story that feels right; all are equally unsupported by evidence.

Scientific explanations can illuminate our situation regarding mortality. Evolution is after all, in essence, a mechanism that develops complex designs capable of persistence. Many of these designs have survived for billions of years and we are composed of a host of these immortal mechanisms.

Mortality enters our picture because we are a product of evolution's replicator-based strategy for persistence: make copies faster than they are destroyed and it will last. Unfortunately, the success of this strategy does not necessarily rely on individual copies persisting for long.

Clearly there are aspects of us that are immortal. As discussed in the previous section, the evolutionary knowledge, of which we are a manifestation, is immortal at least compared to an individual human life. A significant portion of our life is devoted to re-traversing this evolutionary journey. What is mortal is our mature self, which upon leaving the immortal evolutionary realm, emerges into the present where each life is an experimental search for designs that can endure; a search for further evolutionary knowledge. All lives are unique and are composed of a multitude of unique events. Each of these events is a probe of design space, a part of the search for the future. Extremely few paths in design space reach designs sufficiently robust to become part of the future's inheritance. A very small number are Einsteins or Mozarts who happen upon enduring design and gain a type of immortality.

If you desire immortality, science can offer some suggestions.

1) Examine and consider redefining your concept of self to align with those cosmic aspects of yourself that are truly immortal.
2) Have children.
3) Create items of cultural significance that will be part of the future's cultural inheritance.

First examine carefully your concept of self. Who exactly is it that is striving for immortality? If we conceive of our self as the mature agent engaged in an almost certainly futile search, we are indeed mortal. The self, fully identified with the day-to-day struggles and challenges, with transient desires and fears, will not live on. These struggles, challenges, desires and fears are 'one time only' events. There will be little future evidence that these aspects of us ever existed and much less chance that they will be a significant part of the future. On the other hand, if we conceive of our self as a manifestation of the timeless evolutionary process and our mature life as but one attempt by this amazing process to create the future, we have secured a true basis for immortality.

Extending our sense of self may be a path that can move us forward for the problem may really only be that we see ourselves too narrowly. The self by focusing on 'one time only' issues defines itself to be mortal. Extending our concept of self to include our immortal aspects may be all that is required for us to truly gain immortality.

Each hair on our head contains a living cell with our complete biological knowledge contained within the cell in the form of its genetics. If we are to pluck a hair from our head, it will die. In a sense, because the hair contains a copy of our unique biology, it is valid to say we have died. Certainly from the hair's point of view, from the point of view that is totally consumed with the day to day 'one time only' issues of that particular cell, it is true to say 'I have died'. But of course the hair is only one of countless hair cells and only one of hundreds of trillions of other body cells, all containing our complete instructions. From this larger being's point of view, from the point of view of the complete organism, it is true that 'I am still alive'. The view of the hair is the point of view of our self. If we are plucked from life then from our self's point of view it is true to say 'I have died'. But we are part of a family lineage, we share virtually all of our genetics and cultural inheritance with billions of other humans, we share 98% of our genetics with other primates. As Einstein noted it is delusional to conceive of ourselves solely as an isolated, individual, mortal being. There is a point of view from which it is true to say 'I have not died' and this point of view may be reachable through extending our sense of self.

It is not easy. Evolution has burdened us with history from our biological past that encourages us to be supremely focused on immediate 'one time only' events like sex, food and resources. This strategy was the primary strategy employed by our evolutionary ancestors to achieve survival. Fortunately for us they were not distracted with abstract strivings for immortality and every single one of our multitudes of ancestors managed to succeed in surviving at least long enough to reproduce.

Cultural evolution has produced systems of knowledge able to co-exist with the instincts and knowledge inherited from our biological past. In many cases these systems can successfully compete with our biological imperatives and become dominant in our conception of self. Indeed a widely made claim of religion is that it tempers our base nature and can save us from it.

This may be even more possible with a scientific world view. A revolutionary point made in Susan Blackmore's fascinating book <u>The Meme Machine</u> (1999) is the assertion that our conception of self is no more than the meme that has come to dominate our brain. She sees the brain as hardware for which an unlimited variety of memetic software competes for the opportunity to be run. When we think a certain thought it is an instance of a meme playing itself on our hardware. We have the thoughts we have for no other reason than they have competed successfully in securing runtime in our brain. Our concept of self has come to be our brain's dominant meme and the one largely responsible for assigning runtime to itself and other related memes. Blackmore proposes two methods that can assist us in wrestling control of our 'self' from the unconscious competitive processes weighed heavily towards 'one-time-only' concerns. These are science and meditation; science because:

> *of its ideals of truth and seeking evidence. It doesn't always live up to these ideals, but in principle it is capable of destroying any untruthful meme-complex by putting it to the test, by demanding evidence, or by devising an experiment.*

Meditation discourages the insistence of the self and encourages other memes to come forward and present themselves for consideration. Lessening the grip of the self may allow less mortal conceptions of ourselves to arise.

Individual death has always been a dominant concern. Human graves and other artefacts of burial ceremonies provide some of the most ancient evidence of human culture. Our need to honour our dead and mark their passing was a cultural constant long before the rise of the great modern religions and will continue long after they have gone. We have a cosmic relationship with our relatives. Each of our fathers and mothers manufactured chemicals into the sex cells that joined to become our single celled biological beginnings. We were prompted through evolutionary time in our mother's womb and instructed and shaped by our parents and the larger human community after our birth. We share the details of this mystical cosmic journey with our kin and must confront it with their passing. Each passing of a loved one provides us with the opportunity to consider the miracle that has fashioned us from the stuff of the universe, which has prepared us to strut and fret our hour on the stage and then be gone.

My father has prepared detailed instructions for a Christian funeral. At the time, I will of course honour these, but it will seem a lost opportunity. Listening to a cleric recounting yet again the story of Jesus being raised from the dead and providing all men with eternal salvation is banal. Do we as individuals and as a culture exist in such poverty of understanding that we must paper over this great mystery with a fairy tale? Can we not take a little time to honestly confront the cosmic epic of which we and our loved ones are fleeting but magnificent expressions?

Acceptance of the central religious message that 'this is not about you' leads to religious experiences in a number of forms. Once we acknowledge there are greater powers and a wider agenda then our day to day concerns we have laid the groundwork for feelings of awe, humility and belonging in a more fundamental reality. This acknowledgement is a mainstay of religious teaching but it is also fundamental to science. Amongst the scientists who study consciousness there is a growing consensus that the existence of the self is an illusion (Blackmore 2002). The entire content of science describes a miraculous reality more fundamental than that wrapped up in our personal strivings and concerns. Compared to religious views of reality, the one described by Universal Darwinism is true in a beautiful and unrivalled sense.

Appendix A - An example of inference or knowledge updating

Let's explore a simple example to clarify how Bayesian updating works. Let's say we are concerned about the chances of it raining tomorrow. We know that the sky has been clear all day today and we are fairly confident it will not rain tomorrow. However this evening we get some new data, a weather forecast predicting rain tomorrow. What effect should this new data have on our confidence level in our hypothesis that it will not rain tomorrow? Obviously the new data, a weather forecast predicting rain, should diminish our confidence in our hypothesis, but exactly what adjustment to our confidence should be made?

To do the calculation let's follow the terminology of Bayesian Probability and analyze the problem by identifying three variables: H, X and I.

H is our hypothesis; it will not rain tomorrow

X is our relevant prior knowledge; the sky was clear today

I is our new data; a weather report predicting rain tomorrow.

Using Bayesian Probability and given the reliability of historical records concerning our weather we can exactly calculate this adjustment. First we need to quantify three related and available probabilities:

> $P(H|X)$ - Probability it will not rain tomorrow given that it was clear all day today. We can estimate this probability by looking through previous historical weather forecasts and data. Let's say we look at the last 100 clear days and then at each of the subsequent days' weather and we find on 85 of

those subsequent days it did not rain. We can estimate this probability at .85. Before we consider the new data (in the form of the weather forecast) we should assign a .85 probability as our confidence level of it not raining tomorrow.

$P(I|X)$ - Probability of a weather forecast for rain tomorrow given that it was clear today. Again we can estimate this probability by looking through historical weather data and forecasts. Let's say that the probability the weatherman predicted rain for the days subsequent to a clear day is .2.

$P(I|HX)$ - Probability of a weather forecast for rain tomorrow given that the sky is clear today and that it does not rain tomorrow. This is an estimation of the probability of the weatherman making the wrong forecast in this situation (when there is a preceding clear day). Let's say the weatherman is pretty good and we see from the historical data that the probability of him getting the forecast wrong in this situation is .05.

We might note that if the weatherman's prediction has any value the probability described in 3) must be smaller than 2) as 2) is the percentage of times the weatherman predicts rain for days following clear days and 3) imposes the further limiting criteria that his prediction proves to be wrong. Let's say that from historical data we see that he was wrong only 25% of the time (out of the 20 days he forecasted rain the day after a clear day it in fact did not rain on 5 of the subsequent days).

Now we have all the numbers we need to calculate the revised probability of our hypothesis being true in light of this the new data.

P(H|IX) Probability of it not raining tomorrow given that the sky is clear today and the forecast is for rain. This is the confidence we should have in our hypothesis and it is the item we are going to calculate.

P(H|X) Probability it will not rain tomorrow given the sky is clear today =. 85

 P(I|HX) Probability of the weather forecast for rain given that it does not rain tomorrow and that the sky is clear today = .05

P(I|X) Probability of a weather forecast for rain tomorrow given the sky is clear today = .2

The Bayesian formula that relates our correct confidence level to our hypothesis, our prior data and our new data is:

$$P(H|IX) = P(H|X)\frac{P(I|HX)}{P(I|X)}$$

1

Substituting in our numbers:

= .85*.05/.2 =.2125

Given the new data we should now assign only a .2125 probability to our confidence level in the hypothesis that it will not rain tomorrow.

A close look at equation 1 might be warranted. The correct initial confidence level in our hypothesis that it will not rain tomorrow, based on our existing knowledge (that it was clear today), is .85. The effect of our new data (the forecast for rain) is to adjust this confidence level by the ratio of two things we know which concern the relevance of our new data:

The probability of this data (a rain forecast) given our prior knowledge (it was clear all day today). This quantity is the denominator of our 'adjustment' ratio, $P(I|X)$.

The probability of this data (a rain forecast) given both our prior knowledge (it was clear all day today) and our hypothesis being correct (no rain actually occurs on the subsequent day). This quantity is the numerator of our 'adjustment' ratio, $P(I|HX)$.

If the weatherman were a genius and always made correct predictions, the denominator would be .85, the same as the probability that it rains the day after a clear day. The numerator would be zero as the infallible weatherman has never predicted rain for a day when it hasn't rained. Thus the ratio would be zero and our degree of confidence in our hypothesis that it will not rain tomorrow should be zero.

If the weatherman had no skill or was lazy and always predicted rain the next day no matter what then the probability in both the denominator and the numerator of the ratio will be 1 as he always predicts rain regardless of anything else. Thus the ratio will be 1/1 or 1 and our level of confidence in our hypothesis (no rain tomorrow) should remain unchanged after receiving the weather forecast.

Let's say the weatherman is an evil genius and can accurately predict the weather but purposely issues a forecast to the public that is the opposite of what he knows will happen; he predicts it will rain when he knows it will not and he predicts it will not rain when he knows it will. In this case, perhaps surprisingly, once we assign the probabilities we can arrive at certainty about our hypothesis. The probability we should assign in the denominator is tricky. The probability, if the weatherman was infallible and honest, of a prediction of rain after a clear day would be .15 but as he always lies in his forecast the probability in this case is .85. The numerator is easier as on those days in which our hypothesis turns out to be true (it does not rain) the evil weatherman will have predicted rain with probability 1. Thus our 'adjustment' ratio is 1/.85 and when we multiply this by our original confidence level (.85) the result is a new confidence level of 1. We can be certain it will not rain tomorrow (our hypothesis) if the evil weatherman predicts it will rain tomorrow.

We can see in these four examples how our confidence in a hypothesis evolves as new data is received. The new data can be substantially informative as in the first example where the confidence level moved from .85 to .2125; it can be extremely informative when the new data infallibly reveals the actual phenomena as with our infallible weatherman; or it can be completely non-informative as with a forecast by a weatherman who always makes the same prediction.

Thus the updating of our hypothesis as to tomorrows weather results in an increase to our knowledge in proportion to the informative weight of the data received. When information is received that changes the hypothesis' probability, our chances of being surprised by tomorrow's actual weather is reduced, that is, we have more accurate knowledge of tomorrow's weather.

Glossary of Terms

Decoherence	The process by which information is copied from a quantum entity into its environment. This process is also called 'the collapse of the wave function'.
Einselection	A term coined by Wojciech Zurek that is short for 'environment induced superselection'. The processes of selection of the specific eigenvector belonging to the wave function of an entangled quantum entity that specifies the type of information capable of being copied into the environment upon decoherence. Examples of the types of information which may be copied are momentum and postion.
Envariance	A term coined by Wojciech Zurek that is short for 'environment–assisted invariance'. It is the process of assignment of probabilities to the various possible outcome of decoherence. The assignments are equivalent to those of Born's rule which had been required as an axiom of quantum theory prior to Zurek's derivation.
Emergent	Complex systems and patterns arising out of a multiplicity of relatively simple interactions.
Fluctuation theorem	A mathematical theorem that describes how the probability of violations of the 2nd law of thermodynamics becomes exponentially

	small as time or the system size increases.
Inference	The act of drawing a conclusion by deductive reasoning from given facts. The unique mathematical method of performing inference is Bayes' theorem.
Instantiation	To represent (an abstract concept) by a concrete or tangible example.
Isomorphism	An equivalence between mathematical objects. If two objects are isomorphic, then any property that is preserved by an isomorphism and that is true of one of the objects, is also true of the other. If an isomorphism can be found from a relatively unknown part of mathematics into some well studied division of mathematics, where many theorems are already proved, and many methods are already available to find answers, then the function can be used to map whole problems out of unfamiliar territory over to "solid ground" where the problem is easier to understand and work with.
Iterative	An iterative method is a mathematical procedure that generates a sequence of improving approximate solutions for a class of problems
Probability	The objective degree of rational belief, given the evidence.

Index

Bibliography

Adams, Paul. "Hebb and Darwin." *Journal of Theoretical Biology*, 1998: 419-438.

Blackmore, Susan. *The Grand Illusion: Why consciousness only exists when you look for it.* New Scientist June 22, 2002, p 26-29, 2002.

—. *The Meme Machine.* Oxford, UK: Oxford University Press, 1999.

Blume-Kohout, Robin, Ng Hui Khoon, Viola Lorenza, and David Poulin. "Characterizing the Structure of Preserved Information in Quantum Processes." *Physical Review Letters, 100 (3). Art. No. 030501. ISSN 0031-9007,* 2008.

Boyer, Pascal. *Religion Explained.* New York: Basic Books, 2001.

Campbell, Donald. "Evolutionary Epistemology." In *In The philosophy of Karl R. Popper,* by P. A. Schilpp, 412-463. Open Court. , 1974.

Campbell, Donald. "Variation and selective retention in socio-cultural evolution." In *Social change in developing areas: A reinterpretation of evolutionary theory,* by George I. Blanksten and Raymond W. Mack (Eds.), Herbert R. Barringer, 19-49. Cambridge, Mass.: Schenkman, 1965.

Campbell, John. "Bayesian Methods and Universal Darwinism." *BAYESIAN INFERENCE AND MAXIMUM ENTROPY METHODS IN SCIENCE AND ENGINEERING: The 29th International Workshop on Bayesian Inference and Maximum Entropy Methods in*

Science and Engineering. AIP Conference Proceedings, Volume 1193. AIP Conference Proceedings, 2009. 40-47.

Chrisantha Fernando, K. K. Karishma, Eörs Szathmáry. "Copying and Evolution of Neuronal Topology." *PLoS ONE 3,* 2008.

Darwin, Charles. *The Origin of Species.* sixth edition. New York: The New American Library - 1958, 1872.

Dawkins, Richard. *The Blind Watchmaker.* New York: W. W. Norton & Company, Inc. ISBN 0-393-31570-3, 1986.

—. *The Selfish Gene.* Oxford University Press, 1976.

Dennett, Daniel C. *Darwin's Dangerous Idea.* New York: Touchstone Publishing, 1995.

Deutsch, David. "Quantum theory, the Church-Turing principle and the universal quantum computer." *Proceedings of the Royal Society of London A 400 , 97-117.,* 1985: 97-117.

—. *The Fabric of Reality.* London: Penguin Books, 1997.

Dictionary.com. "http://dictionary.reference.com/browse/faith." As viewed May 9, 2010.

Dishoeck, E.F. van. "The chemistry of diffuse and dark interstellar clouds." In *The Molecular Astrophysics of Stars and Galaxies,* by T.W. Hartquist and D.A. Williams. Oxford: Oxford University Press, 1998.

Dobzhansky, Theodosius. "Nothing In Biology Makes Sense Except In The Light of Evolution." *American Biology Teacher, vol. 35,* 1973: 125-129.

Edelman, Gerald. *Neural Darwinism: The Theory of Neuronal Group Selection.* Basic Books, 1987.

Einstein, Albert. *Einstein's 1954 letter on religion.* As viewed September 28, 2010 http://www.relativitybook.com/resources/Einstein_religio n.html, 1954.

Einstein, Albert. *Science and Religion.* New York Times magazine, 1930.

Einstein, Albert. "What I believe." 1930.

Emes, RD, AJ Pocklington, CN Anderson, A Bayes, and MO Collins. "Evolutionary expansion and anatomical specialization of synapse proteome complexity." *Nat Neurosci.,* 2008: 799–806.

Feyman, Richard. As viewed on Wiki Quotes, 11-14-08. http://en.wikiquote.org/wiki/Richard_Feynman, 2009.

Ford, Kenneth. *John Archibald Wheeler: Doer and Visionary.* As viewed on http://www.metanexus.net/magazine/ArticleDetail/tabid/ 68/id/5491/Default.aspx September 24, 2010, 2010.

Friston, Karl, and Stephan Klass. "Free Energy and the brain." *Synthese, 159,* 2007: 417-458.

Gilbert, Walter. "The RNA World ." *Nature 319: 618. doi:10.1038/319618a0,* 1986.

Gould, Stephen Jay. *Rocks of Ages: Science and Religion in the Fullness of Life.* Ballantine Books, ISBN 0-345-43009-3, 1999.

Grabowsk, Laura M., Wesley R. Elsberry, Charles Ofria, and Robert T. Pennock:. " On the evolution of motility and intelligent tactic response." *GECCO*, 2008: 209-216.

Green, Brian. *The Elegant Universe.* New York: W.W. Norton & Company Inc., 1999.

Gregory, Richard. " Brainy Mind." *Brit. Med. Journal* , 1998: 1693 - 5.

Guth, Alan H. "Inflation." In *Measuring and Modeling the Universe,* by W.L. Freedman (Editor). Cambridge: Cambridge University Press, 2004.

Hamma, Alioscia, Fontini Markopoulou, Isabeau Premont-Schwarz, and Simone Severini. "Lieb-Robinson bounds and the speed of light from topological order." *Journal-ref: Phys.Rev.Lett.102:017204,2009,* 2009.

Hebb, Donald O. *The Organization of Behavior: A neuropsychological theory.* New York: Wiley, 1949.

Henrich, J. and R. McElreath. "Dual Inheritance Theory: The Evolution of Human Cultural Capacities and Cultural Evolution." In *Oxford Handbook of Evolutionary Psychology,* by edited by Robin Dunbar and Louise Barrett. Oxford University Press, 2007.

Henrich, J. and R. McElreath. "Dual Inheritance Theory: The Evolution of Human Cultural Capacities and Cultural Evolution."

Holldobler, Bert, and E.O. Wilson. *Superorganism.* WW Norton, 2008.

Huang, Gregory T. "Is This a Unified Theory of the Brain." *New Scientist,* 2008, May: 30-33.

Iacoboni, M. "Understanding others: imitation, language, empathy." In *Perspectives on imitation: from cognitive neuroscience to social science*, by S. Hurley and N. Chater. Cambridge, MA: MIT Press, 2005.

Irving, Klotz. *Postmodernist Rhetoric Does Not Change Fundamental Scientific Facts,.* website http://hps.elte.hu/~gk/Sokal/Sokal/KLotz.html, as viewed January 5, 2009, 2009.

Isaacson, Walter. *Einstein: His Life and Universe.* Simon and Schuster, ISBN-10: 1416586911, 2008.

Jaynes, Edwin T. "Bayesian Methods: General Background." In *Maximum-Entropy and Bayesian Methods in Applied Statistics*, by J. H. Justice (ed.). Cambridge: Cambridge Univ. Press, 1986.

Jaynes, Edwin T. "Gibbs vs Boltzmann Entropies." *Am. J. Phys. , 391.*, 1965.

—. *Probability Theory: The Logic of Science.* University of Cambridge Press, 2003.

Jaynes, Edwin T. "Where do we Stand on Maximum Entropy?" In *The Maximum Entropy Formalism*, by R. D. Levine and M. Tribus. Cambridge, MA: M. I. T. Press, 1979.

Kauffman, Louis H. "The Mathematics of Charles Sanders Peirce." *Cybernetics & Human Knowing, Vol.8, no.1–2*, 2001: 79–110.

Kirby, S and Christiansen, M. *Language Evolution.* New York: Oxford University Press, 2003.

Lewontin, Richard C. *The Genetic Basis of Evolutionary Change.* Columbia University Press, 1974.

Lloyd, Seth. *Programming the Universe*. Vintage; Reprint edition, 2007.

Misner, Charles, Kip Thorne, and John Archibald Wheeler. *Gravitation*. W.H. Freeman and Company, 1973.

O'Brien, M. J., and R. L. Lyman. "Resolving Phylogeny: Evolutionary Archaeology's Fundamental Issue." In *Essential Tensions in Archaeological Method and Theory,*, by T. L. VanPool and C. S. VanPool, 115-135. Salt Lake City: University of Utah Press, 2003.

Online, Merriam Webster. "As viewed on http://www.merriam-webster.com/dictionary/objectivity." 2008.

Plotkin, Henry. *Darwin Machines*. Cambridge Massachusetts: Harvard University Press, 1993.

Popper, Karl. *Objective Knowledge* . Clarendon Press, 1972.

Prigogine, Ilya. *Order Out Of Chaos*. Bantam; Reissue edition (Mar 1 1984), 1984.

Ricklefs, Robert E. *Ecology*. Concord, Massachusetts: Chiron Press, 1979.

Rundle, Bede. *Why the is something rather than nothing*. Oxford University Press, 2004.

Sension, Roseanne. "Biophysics: Quantum paths to photosynthesis." *Nature 446, 740-741*, 2007: 740-741.

Shear, Johnathan. *Explaining Consciousness: The "Hard Problem"*. The MIT Press, 1999.

Simpson, James. "James Simpson Quotations: http://www.bartleby.com/63/15/3115.html; Last viewed September 4, 2004." 2004.

Smolin, Lee. "Newtonian Gravity in Loop Quantum Gravity." *arXiv:1001.3668v2 [gr-qc]*, 2010: 16.

Smolin, Lee. "The case for background independece." *arXiv:hep-th/0507235v1*, 2005.

Spencer-Brown, G. *The Laws of Form*. New York: E.P. Dutton, 1979.

Stenger, V.J. *The Comprehensible Cosmos*. New York: Prometheus Books, 2006.

Tegmark, Max. "The Mathematical Universe." *Foundations of Physics 38*, 2009: 101-50. .

Verlinde, Erik. "On the Origin of Gravity and the Laws of Newton." *arXiv:1001.0785*, 2010.

Wigner, Eugene. *Symmetries and Reflections: Scientific Essays*. MIT Press. ISBN 0-262-73021-9, 1970.

Wigner, Eugene. "The Unreasonable Effectiveness of Mathematics in the Natural Sciences." *Communications on Pure and Applied Mathematics 13*, 1960: 1–14.

Wikipedia. *Bayesian Probability*. http://en.wikipedia.org/wiki/Bayesian_probability, as viewed January 6, 2008, 2008.

Wikipedia. "http://en.wikipedia.org/wiki/Culture."

Wikipedia. "Population Genetics." http://en.wikipedia.org/wiki/Population_genetics, as viewed Sept. 11, 2010.

Zurek, Wojciech H. "Decoherence, einselection and the existential interpretation (the rough guide)." *Philosophic Transactions of the Royal Society; vol. 356 no. 1743*, 1998: 1793-1821.

Zurek, Wojciech H. "Decoherence, einselection and the quantum origins of the classical." *Rev. Mod. Phys. 75, 715*, 2003.

Zurek, Wojciech H. "Quantum Darwinism." *Nature Physics, vol. 5*, 2009: 181-188.

Zurek, Wojciech H. "Quantum Darwinism and envariance." In *Science and Ultimate Reality: From the Quantum to the Cosmos*, by P. C. W. Davies, and C. H. Harper, eds. J. D. Barrow. Cambridge University Press, 2004.

Zurek, Wojciech H. "Relative States and the Environment: Einselection, Envariance, Quantum Darwinism and the Existential Interpretation." *arXiv:0707.2832v1, http://arxiv.org/PS_cache/arxiv/pdf/0707/0707.2832v1.pdf*, 2007.

Zwolak, M., H.T. Quan, and W.H. Zurek. "Quantum Darwinism in non-ideal environments." *Preprint; arXiv:0911.4307*, 2009.

www.ingramcontent.com/pod-product-compliance
Lightning Source LLC
Chambersburg PA
CBHW071422170526
45165CB00001B/359